Geological Factors and the Evolution of Plants

Geological Factors and the Evolution of Plants

Edited by Bruce H. Tiffney

Yale University Press
New Haven and London

Library of Congress Cataloging in Publication Data
Main entry under title:
Geological factors and the evolution of plants.
   Includes index.
    1. Paleobotany — Congresses.   2. Geology — Congresses.
I. Tiffney, Bruce H.
QE901.G45   1985   561   84–27053
ISBN 0–300–03304–4

Designed by James J. Johnson
and set in Times Roman type.
Printed in the United States of America by
Vail-Ballou Press, Binghamton, New York.

10   9   8   7   6   5   4   3   2   1

Contents

Preface

The idea for the symposium that prompted this book arose in a dormitory room in 1979 in Stillwater, Oklahoma, at the time of the annual meeting of the American Institute of Biological Sciences. Several colleagues had gathered to review progress in the field of paleobotany in the most useful colloquium of any national meeting — the after-hours bull session. Attention turned to the question of what topic (if any) the Paleobotanical Section of the Botanical Society of America might wish to sponsor at the upcoming Third North American Paleontological Convention. Discussion went around the room, each individual suggesting a theme or important topic appropriate for a symposium of general interest to paleontologists. When my turn came, I suggested the synergistic interaction of organism and environment as viewed on the paleontological time scale. This idea ultimately caught the imagination of those present and launched this assembly of papers.

It was agreed from the outset that we wished to encourage publication of the ideas that a symposium on this subject would attract, so a symposium "committee" was constituted. In honor of having suggested the idea, I was promptly "elected" chairman. This group aided in the planning of the symposium and then served as an editorial board, from which two reviews for each written submission were obtained. Without the assistance of these colleagues, my job as chief organizer would have been immensely more difficult and time-consuming; I wish to thank Patricia Gensel (University of North Carolina), Leo Hickey (Yale University), Andrew Knoll (Harvard

University), Karl Niklas (Cornell University), and Tom Phillips (University of Illinois) for their unflagging enthusiasm, assistance, and support throughout this undertaking. Of course, as the chairman and final editor of this collection, I take full responsibility for any oversights, errors, or omissions that escaped their eyes or mine. A special note of appreciation is due Patricia Gensel and Andrew Knoll, who co-chaired the symposium at Montreal during my unanticipated absence by virtue of an opportunity to spend August 1982 in western China.

I must also express my gratitude to Edward Tripp of Yale University Press, who has guided me in the preparation of the submitted manuscripts for publication, making a time-consuming job more efficient and bearable.

Bruce H. Tiffney

BRUCE H. TIFFNEY

Geological Factors and the Evolution of Plants

Paleontology is the most important source of information on evolution; it provides significant data against which all evolutionary hypotheses must be tested. However, the fossils studied by paleontologists are but a minuscule subset of all life that lived in the past. This disparity has led to the general conception of the incomplete and thus fallible nature of the fossil record. As a result, paleontologists long have tended to concentrate on the accumulation of raw data through the description of species and to shun theorization as inappropriate in light of the supposed absence of sufficient information.

More recently, paleontologists have come to appreciate both the importance of theory as a tool for the organization of data and the significant contribution that the fossil record makes to our understanding of evolution. Synthetic investigations of pattern and process based on the fossil record have become more common, first among vertebrate paleontologists and subsequently involving all of paleontology. In general, these inquiries have focused on lineages, treating either the patterns and processes of evolution within a lineage (for example, Knoll, Niklas, Gensel, and Tiffney, 1984) or the patterns of features that emerge from the comparison or cumulative examination of separate lineages (Sepkoski, 1978, 1979; Niklas, Tiffney, and Knoll, 1980, 1983). However, all evolution, including that of lineages, occurs within the constraints established by the synergistic evolution of the physical environment of the earth and its resident biota. Significant as this interaction is, it has been touched upon only rarely in paleontology and

1

has not been addressed in a systematic manner. Cognizance of the importance of this interaction has grown in the past decade. The recognition that the same process may yield radically different results depending upon the rate at which it occurs led to the realization that past and present environments may have been remarkably different even though they were influenced by the same physical processes. From this realization arose the perception that certain periods of earth history possessed unique attributes (the coal of the Carboniferous, the chalk of the Cretaceous). This observation raises the challenge that the timing of biotic events (the origin of eukcaryotes, the floral invasion of land) may have been governed by unique environmental factors.

By raising the question of the synergism of organism and environment in the title of the symposium on which this book is based, the sponsors sought to stimulate the investigation of basic interactions of biota and planet and thus to advance the search for process in evolution. A paleobotanical approach to this question is not inappropriate; many marine and most terrestrial plants are sedentary, so they must tolerate or modify the existing environment in order to survive. Recognition of the effect of biota on environment is reflected in the concept of succession as applied to modern plant communities, wherein successive ecological groupings of plants modify their physical environment such as to ensure their own local disappearance and pave the way for the expansion of the next ecological group. In an analogous manner, the evolution of the terrestrial flora (and its associated fauna) may be seen as a slow-motion succession in which earlier groups modified the environment such as to pave the way for the evolutionary appearance of their successors. The question remains, what determines the timing of change in the system? Is the succession analogy just that (and do extrinsic factors determine the timing), or is the analogy more of a homology than it appears on the surface?

With such a broad question to address and all of earthly time, space, and biota to choose from for topics of investigation, it is no wonder that the chapters of this book are disparate in nature: Although individual contributions address different aspects of this question at different levels of complexity and with regard to different times in earth history, they all contain the common thread of some aspect of the synergistic interaction of planet and biota — some recording the influence of physical environment on biota and some the reverse. No author takes synergism through a "full cycle" (physical-biological-physical), but this pattern is implicit in this book and in other aspects of earth history not addressed here. Inasmuch as these chapters provide a limited survey of the total question, I will introduce them in the context of a personal view of some of the salient events in the history of synergism of organisms and environment.

Perhaps this synergism has been best documented in studies of the Precambrian. Initially, the synthesis and survival of organic molecules depended upon energy and materials derived from the surface and atmosphere of the planet and took place in a reducing environment. Through repeated evolutionary experiments, autotrophic organisms (presumably blue-green algae) joined, and ultimately came to support, the heterotrophic communities of the early Precambrian. These algae liberated oxygen as a function of their metabolism and thereby threatened to alter the entire chemistry of the planet. However, their productivity was apparently limited by the absence of a stable habitat suitable to support them in large numbers. This lack of habitat appears to have been due to the highly active tectonic environment of the Archean eon (Knoll, 1983). It was only with the reduction of tectonic activity in the early Proterozoic eon that sizable, stable near-shore environments appeared for the first time, permitting oxygen-producing autotrophs to occur in sufficient quantity for oxygen production to exceed oxygen consumption — and therefore for an oxygen-rich atmosphere to develop. The consequences of this change — which altered the whole chemistry of the earth's surface and its atmosphere and the options open to the subsequent evolution of life — were immense. The increased diversity of available habitat was likely reflected in greater diversity of both auto- and heterotrophs. The increased biotic interactions resulting from this diversity could be expected to have speeded up the pace of evolution, an example perhaps being the origin of eukaryotes through endosymbiosis. This provides an illustration of a full cycle of change involving a biotic influence (the production of oxygen) on the physical environment that in turn influenced biotic evolution. The timing and environmental milieu of the evolution of eukaryotes is examined by Stanley M. Awramik and James W. Valentine in their chapter, "Adaptive Aspects of the Origin of Autotrophic Eukaryotes." Similar questions could be asked of environmental influences on the time of appearance of metazoans. Was their evolution a function of environmental stimuli, or was it a direct result of the pace of biological diversification? Similarly, which factor (if either) influenced the appearance of hard parts and the burst of invertebrate diversification witnessed at the Precambrian/Cambrian boundary?

From the botanical perspective, the next major event in the history of life was the invasion of the land in the Silurian and Devonian. Was the appearance of a land flora simply a function of the pace of biotic differentiation (thus logically following in due time from the appearance of eukaryotes), or was it mediated by external environmental effects? The former is virtually unexaminable as a hypothesis, but the geologically unique aspects of this time are open to investigation. As an example, Ordovician glaciations (Windley, 1977) may have caused a decrease in sea level and thereby stressed near-shore communities. This in turn could have led to the evolution of stress-tolerant intertidal and subtidal algae adapted to a degree of dessication and having the potential to explore terrestrial environments (Pratt, Phillips, and Dennison, 1978). In this vein, David Chapman explores the importance of oxygen for land-plant metabolism in his chapter, "Geological Aspects and Biochemical Aspects of the Origin of Land Plants." He concludes that the partial pressure of oxygen in the earth's atmosphere could have directly influenced the timing of the invasion of the land. Again, the atmospheric composition created by prokaryotes in the Precambrian might prove to be the factor that ultimately permitted the establishment of the terrestrial ecosystem. We know little of the full spectrum of possible constraints on this process; they may be inferred from the biology of living organisms (as Chapman has done), or they may be inferred directly from patterns in the geologic record (for example, from the glaciations mentioned above) or from the fossil record. Richard Beerbower pursues the latter route in his chapter, "Early Development of Continental Ecosystems," wherein he reviews the status of the freshwater and terrestrial ecosystems of the Silurian and Devonian in light of the coexisting near-shore marine communities from which the terrestrial biota must have arisen. His coverage is eclectic, and he touches on many of the questions that I raise here as well as on some aspects of Chapman's chapter. Although Beerbower does not suggest a specific environmental or biological trigger of this invasion, he summarizes the paleontological milieu in which it occurred and adduces a model for the early evolution of terrestrial communities based on the ecological factors involved in the transition from marine to freshwater to terrestrial ecosystems.

The appearance of land flora is paleobotanically significant, but it also had an equally important effect on the evolution of terrestrial and marine systems. On land, the existence of a significant biomass of plants established the basis for a terrestrial food chain. Arthropods were apparently the first terrestrial herbivores (Kevan, Chaloner, and Savile, 1975; Stormer, 1977), and one would think that they would be followed by carnivores and other herbivores. However, the evolutionary history of the late Paleozoic terrestrial food web has not been fully elucidated (Scott, 1980), and it is unclear when a food pyramid that included terrestrially based vertebrate herbivores and carnivores was fully established; it may not have appeared until the late Pennsylvanian or Permian (Panchen, 1973; Scott, 1980). It is surprising that the well-developed terrestrial vegetation did not attract a diverse herbivorous biota before this time. Why did it take so long to establish a complex terrestrial food chain, and what was the pattern of its establishment? Was it governed by the internal dynamics of the ecosystem or by extrinsic physical factors?

Turning to effects on the physical environment, one must consider that the introduction of plants to land must have had a tremendous influence on the chemical composition and sedimentary regimes of fresh and marine waters. Land plants secrete substances that speed up weathering and presumably alter the amounts and kinds of chemicals entering the atmosphere and oceans. They also physically break up rock materials. Concomitantly, they serve to retard rates of erosion by binding the soil with their roots and rhizomes. Finally, they represent a chemically distinct source of carbon in the biosphere. Preliminary evidence suggests that these factors were not without their effect; sulfur ratios in marine sediments indicate that the appearance of land plants corresponds with an increase in refractory organic matter in the marine environment (Berner and Raiswell, 1983). This organic matter is apparently derived from the lignin-rich residue of land plants, which degraded more slowly than carbon of marine origin. As a result, marine sediment-feeding organisms had to burrow more deeply into the mud in order to find available carbon. Thus the appearance of the land flora is correlated with the development of a marine infauna that fed at greater depths beneath the sediment surface than had the previous infauna. The degree to which rates of carbon burial and degradation were affected by changes in sediment influx from terrestrial environments is unclear.

The advance to land involved significant adaptations to the physiological drought of terrestrial environments, but even with these adaptations land-plant distributions have been decisively influenced by global climates and continental positions. Climate and geography are not independent, as the latter has a strong influence on atmospheric and oceanic circulation and thus on the global distribution of moisture and heat. The effect of geography and paleoclimate on paleophytogeography has long been recognized, but only recently have geophysical techniques provided an independent source of paleogeographic and paleoclimatic reconstructions (see, for example, Scotese et al., 1979; Bambach, Scotese, and Ziegler, 1980; Parrish and Curtis, 1982). Although these reconstructions are first approximations requiring refinement, they provide an independent model against which to check biogeographic data and from which to search for stimuli that would explain observed evolutionary events. Stephen Barrett's chapter, "Early Devonian Continental Positions and Climate: A Framework for Paleophytogeography," provides the geophysical background for the period of time during which the terrestrial flora was just becoming firmly established; the distribution of this flora is explored in this context by Anne Raymond, William Parker, and Stephen Barrett in their chapter, "Early Devonian Phytogeography." Paleogeographic reconstructions of the Carboniferous have been suggested by several authors, and Anne Raymond, William Parker, and Judith Tottman Parrish explore the "Phytogeography and Paleoclimate of the Early Carboniferous" based on the continental reconstructions provided by Bambach, Scotese, and Ziegler (1980).

Trends in global phytogeography are perhaps the largest-scale and most obvious result of continental movement. On a more local scale, however, changing climate and physiography also influence distribution and evolution. Within the Carboniferous coal swamps, lineages evolved and went extinct, often in response to changing habitats caused by climatic and orographic alterations. William DiMichele, Tom L. Phillips, and Russell Peppers trace such a pattern in "The Influence of Climate and Depositional Environment on the Distribution and Evolution of Pennsylvanian Coal-Swamp Plants." The great Carboniferous coal seams attest to the substantial biomass of these communities and hint at the possible influences of this carbon sink on world carbon supplies and on the gaseous composition of the atmosphere. For the first time, sufficient carbon was present in the terrestrial system to support large-scale combustion. Further, the same organisms that fixed the carbon may have raised oxygen levels to a point that would permit combustion. The question of the composition of the Carboniferous atmosphere and the potential for forest fire in the coal swamps is addressed by Michael Cope and William Chaloner in "Wildfire: An Interaction of Biological and Physical Processes." The effect of the fixation of carbon in coal-swamp communities upon the world carbon balance, which has not been explored in a paleontological context, remains moot.

By the late Carboniferous the earth's continents were coalescing to form Pangaea. The effects of spreading continental climates and upland environments, first seen in the mid Pennsylvanian (Phillips et al., 1974), culminated in the extinctions of the Permo-Triassic, which were as significant on land (Niklas, Tiffney, and Knoll, 1980, n.d.) as on sea (Raup and Sepkoski, 1982). The Permo-Triassic extinction is perhaps one of the most striking examples of the direct influence of physical environment upon terrestrial plant communities; it brought about the disappearance of many lineages and presumably the alteration of invertebrate and vertebrate terrestrial food chains. On the physical side, the demise of the swamps removed major sediment traps from between eroding uplands and near-shore marine environments. Perhaps this influenced extinctions in the marine realm (consider the effect of increased turbidity on corals and other filter feeders in the present day). As the lowland communities shrank, upland communities expanded to become the world flora. In contrast to the pteridophyte-dominated swamp vegetation of the Carboniferous, the new plants formed an apparently monotonous vegetation of conifers, cycads, cycadeoids, and other, less well understood gymnosperms (Winston, 1983), with a limited ground cover of ferns and their allies in damper sites. The general biology of modern gymnosperms leads one to suspect that early Mesozoic communities were not particularly efficient at preventing erosion. The sediment that flowed into the aquatic realm was perhaps accompanied by increased amounts of phenols and tannins, the principal protective chemicals of gymnosperms. From what we know of modern gymnosperms, one also suspects that the amount of primary productivity was low, perhaps lower than the late Paleozoic level. Low productivity may have affected marine ecosystems and the atmosphere, and it certainly influenced terrestrial communities. Terrestrial vertebrate associations of the Mesozoic are generally envisioned as analogous to those of the present, but is this necessarily so? Higher nutrient loss through erosion and the lower rate of primary productivity inherent in gymnosperm physiology suggest either (1) that total herbivore and carnivore biomass might have been notably less than that of the present day or (2) that the physiology of these animals functioned on a lower caloric intake than that of modern endotherms (Weaver, 1983).

The evolution of angiosperms must have had a profound effect upon the physical and biological environment. The influence of the group on the evolution of insects, birds, and mammals is well appreciated, but what of the consequence of angiosperm evolution on the marine realm and on the physical environment? Certainly the group must have had a tremendous influence on erosion and sedimentation. Rhizomatous growth is common in many dicots and dominates the monocots in the present day. One can only assume that, from their first appearance along early Cretaceous streambanks (Doyle and Hickey, 1976; Hickey and Doyle, 1977), the angiosperms strongly retarded erosion. In this respect, perhaps one of the most important geological aspects of the evolution of the group occurred in the Oligocene and Miocene, when the grasses diversified in times of increasingly arid world climates. We have no measure of the ability of preceding communities to forestall erosion in seasonally dry environments, but the effectiveness of grasses at controlling erosion is so unique that one can only surmise that the spread of grasslands had a significant effect on the influx of sediment into freshwater and marine systems. As a result, I would suspect that this change in erosion rate should be observable in the evolutionary patterns of near-shore filter-feeding invertebrates and in the sedimentary record in the later Cretaceous and Tertiary. Angiosperms are also the premier chemists of the plant world; some have even been thought to modify their environment such as to reduce or eliminate competition. One can only wonder about the effect of millions of years of leaching of angiosperm chemical byproducts on the chemistry of fresh and marine waters and on their biota. A similar but distinct chemical influence might be exerted by the grasses, which evolved a coating of silicified epidermal cells in response to consumption by horses and other grazing animals. The concentrated silica is released upon the plant's death and enters the sedimentary environment in a finely particulate and easily dissolved form. What effect has this had on freshwater or near-shore marine communities?

The foregoing has been a rather eclectic tour of "what-ifs" in the evolution of the biosphere, with an emphasis on the history of land plants. I neither pose all the questions nor explore all the ramifications of the questions that I do pose. Rather, it has been my intent to establish a perspective. Biologists, geologists, and paleontologists now have both the necessary volume of facts and theoretical framework to begin an exploration of the synergistic interactions that underlie the evolution of the biosphere and that define its unique and unidirectional history. The chapters of this book address greater and lesser aspects of that synergism. I hope that, between the perspective I have presented here and the specific contributions that follow, this volume will stimulate further research on the larger patterns linking the organisms and systems that we normally study separately. The facts that paleontologists have so diligently labored to accumulate will take on full significance only when examined in light of the larger patterns to which they contribute. Those larger patterns are the focus of this book.

References

Bambach, R. K., C. R. Scotese, and A. M. Ziegler. 1980. Before Pangea: The geographies of the Paleozoic world. *Am. Sci.* 68:26–38.

Berner, R., and R. Raiswell. 1983. Burial of organic carbon and pyrite sulfur in sediments over Phanerozoic time: A new theory. *Geochim. Cosmochim. Acta* 47:855–62.

Doyle, J. A., and L. J. Hickey, 1976. Pollen and leaves from the mid-Cretaceous Potomac group and their bearing on early angiosperm evolution. In *Origin and early evolution of angiosperms*, ed. C. B. Beck, 139–206. New York: Columbia Univ. Press.

Hickey, L. J., and J. A. Doyle. 1977. Early Cretaceous fossil evidence for angiosperm evolution. *Bot. Rev.* 43:3–104.

Kevan, P. G., W. G. Chaloner, and D. B. O. Savile. 1975. Interrelationships of early terrestrial arthropods and plants. *Palaeontology* 18:391–417.

Knoll, A. H. 1983. Tectonics, productivity, and ecosystems on the early earth. *Proc. Third North American Paleontological Conf.* 2:307–11.

Knoll, A. H., K. J. Niklas, P. G. Gensel, and B. H. Tiffney. 1984. Character diversification and patterns of evolution in early vascular plants. *Paleobiology* 10: 34–47.

Niklas, K. J., B. H. Tiffney, and A. H. Knoll. 1980. Apparent changes in the diversity of fossil plants. In *Evolutionary biology*, ed. M. K. Hecht, W. C. Steere, and E. Wallace, 12:1–89. New York: Plenum.

———. 1983. Patterns in vascular land plant diversification. *Nature* 303:614–16.

———. In press. Patterns in vascular land plant diversification: A factor analysis at the species level. In *Factors in Phanerozoic diversity*, ed. J. W. Valentine. Princeton: Princeton Univ. Press, forthcoming.

Panchen, A. L. 1973. Carboniferous tetrapods. In *Atlas of palaeobiogeography*, ed. A. Hallam, 117–25. Amsterdam: Elsevier.

Parrish, J. T., and R. L. Curtis. 1982. Atmospheric circulation, upwelling, and organic-rich rocks in the Mesozoic and Cenozoic eras. *Palaeogeogr., Palaeoclimatol., Palaeoecol.* 40:31–66.

Phillips, T. L., R. A. Peppers, M. J. Avcin, and P. F. Laughnan. 1974. Fossil plants and coal: Patterns and change in Pennsylvanian coal swamps of the Illinois basin. *Science* 154:1367–69.

Pratt, L. M., T. L. Phillips, and J. M. Dennison. 1978. Evidence of non-vascular land plants from the early Silurian (Llandoverian) of Virginia, U.S.A. *Rev. Palaeobot. Palynol.* 25:121–50.

Raup, D. M., and J. J. Sepkoski. 1982. Mass extinctions in the marine fossil record. *Science* 215:1501–03.

Scotese, C. R., R. K. Bambach, C. Barton, R. Van der Voo, and A. M. Ziegler. 1979. Paleozoic base maps. *J. Geol.* 87:217–68.

Scott, A. C. 1980. The ecology of some upper Paleozoic floras. In *The Terrestrial environment and the origin of land vertebrates*, ed. A. L. Panchen, 87–115. Systematics Association Special Volume 15. New York: Academic Press.

Sepkoski, J. J., Jr. 1978. A kinetic model of Phanerozoic taxonomic diversity: I. Analysis of marine orders. *Paleobiology* 4:223–51.

———. 1979. A kinetic model of Phanerozoic taxonomic diversity: II. Early Phanerozoic families and multiple equilibria. *Paleobiology* 5:222–51.

Stormer, L. 1977. Arthropod invasion of land during late Silurian and Devonian times. *Science* 197:1362–64.

Weaver, J. C. 1983. The improbable endotherm: The energetics of the sauropod dinosaur *Brachiosaurus*. *Paleobiology* 9:173–82.

Windley, B. J. 1977. *The evolving continents*. New York: J. Wiley and Sons.

Winston, R. B. 1983. A late Pennsylvanian upland flora in Kansas — Systematics and evolutionary implications. *Rev. Palaeobot. Palynol.* 40:5–32.

STANLEY M. AWRAMIK AND
JAMES W. VALENTINE

Adaptive Aspects of the Origin of Autotrophic Eukaryotes

"Only the eucaryotic cell appears ... to have contained the potentialities for the development of highly differential multicellular biological systems, and accordingly only this kind of cell was perpetuated in the evolutionary lines which eventually gave rise to higher plants and animals."

R. Y. Stanier and C. B. von Neil (1962, p. 33)

Introduction

The origin of eukaryotes is sometimes regarded as the most important of all evolutionary steps (Schopf and Oehler, 1976), since it is from eukaryotes that all multicellular kingdoms of organisms have descended. However, important evolutionary questions about this event remain unanswered. Here we examine four questions that deal with adaptive aspects of eukaryotic origins: (1) What adaptive features in the ancestors of eukaryotes favored or permitted the rise of eukaryotic organization? (2) What adaptive opportunities were associated with the origin of eukaryotes? (3) What features of the eukaryotes permitted them to give rise to whole kingdoms of multicellular organisms while in contrast the prokaryotes have never achieved much complexity? (4) Were there any changes on the Precambrian Earth that promoted the evolution of eukaryotes? We propose a generalized answer that speaks to all these aspects of eukaryotic evolution. The differences between eukaryote and prokaryote cells have been reviewed elsewhere (Cavalier-Smith, 1981; Starr and Schmidt, 1981) so it is only necessary to highlight some of the major features that bear on our thesis.

The Origins of Some Autotrophic Eukaryote Grades

Eukaryotes that have achieved relatively large body sizes, such as some algae and all vascular plants and animals, are multicellular. *Multicellu-*

larity is one of those terms that is widely used yet seldom defined. Prokaryotes exhibit a rudimentary multicellularity, "in which there is (i) a plurality of cells, with (ii) a persistent and characteristic sort of juxtaposition, and (iii) in which there is a discernible distinction in structure, function or both among the cells comprising the populations" (Starr and Schmidt, 1981, 11). The fruiting body of myxobacteria is a good example of this (Parish, 1979). However, even these "multicellular" prokaryotes are small; some are barely visible with the unaided eye. It is only in the eukaryotes that large, macroscopic, multicellular organisms with pronounced cellular differentiation and structurally elaborate cell junctions are found. Taylor (1974) reserved the term *multicellular* for organisms comprised of multiple polygenomic units (as opposed to pluricellular prokaryotes, which are composed of multiples of monogenomic units). Thus, only eukaryotes can be truly multicellular. Clearly, this definition of multicellularity begs the question as to why prokaryotes are not multicellular by excluding them for an ad hoc reason. Nevertheless, Taylor is probably on the right track in identifying genome structure as a significant factor in multicellularity.

Multicellular organisms are not only large but complex, and they exhibit a vast range of morphology. It is thought that this range arises primarily from variation in patterns of gene expression because the organizational differences among the genomes of multicellular forms appear to be more important than the biochemical differences (see, for example, Britten and Davidson, 1969, 1971; Campbell, 1982; Hunkapiller et al., 1982). The versatility and sophistication of eukaryotic regulatory systems are evident from the most cursory survey of organismal morphologies (Valentine and Campbell, 1975). Presumably, the development of the present morphologically diverse eukaryotic biota must have chiefly involved the evolutionary genetics of regulatory systems.

Prokaryotes stand in stark contrast. They display a very broad range of metabolic systems, some of which are unknown in eukaryotes; for example, anoxygenic photosynthesis, chemolithotrophy, and methanogenesis (Carlile, 1982). But prokaryotes have achieved little visible morphological complexity; perhaps the branching filamentous cyanobacteria (Geitler, 1932) and fruiting myxobacteria (Parish, 1979) represent the most complex prokaryotic morphologies. Prokaryotes nevertheless perform at least 90 percent of the enzymatic functions of eukaryotes and resemble them closely biochemically (Britten and Davidson, 1971). Some of the gene regulatory systems of prokaryotes are well known and, so far as can be told, they are much simpler than those of eukaryotes. Indeed, our first hypothesis is that the failure of prokaryotes to produce morphologically complex large organisms via cell multiplication and differentiation is due simply to their lack of sufficient gene regulatory sophistication. Once the appropriate advances in regulatory abilities were made, multicellularity became a relatively easy evolutionary step to take; there seem to have been at least 17 separate multicellular inventions from unicellular eukaryotes (Stebbins in Dobzhansky et al., 1977, 395) and probably many more.

If our hypothesis is true, it leads to interesting notions on the origin of eukaryotes. A leading scenario to account for the origin of major eukaryotic organelles — major components of eukaryotic complexity — is that the organelles descended from a series of originally free-living prokaryotes that had become symbionts within an ingesting prokaryotic host cell (Schimper, 1883; Mereschkovsky, 1905; Margulis, 1970). The symbionts are inferred to have lost their independence and to have become integrated into the host as specialized organelles. Although there are problems with this interpretation (Doolittle, 1980; Trench, 1982) and competing theories for the origin of eukaryotes (Bogorad, 1975; Raff and Mahler, 1975), the serial endosymbiosis theory seems successful in accounting for the origin of plastids and probably mitochondria (Cavalier-Smith, 1981) and has been taken to imply that, in some ways, the eukaryotic cell may be regarded as a "multicellular prokaryote" (Taylor, 1974).

This leads to our next hypothesis: that the rise of the eukaryote grade involved the origin of a gene regulatory system, more sophisticated than preceding prokaryotic systems, that was able to integrate the symbiotic organelles so as to function harmoniously within a relatively complex unicellular ground plan. Just as multicellular eukaryotes (fungi, animals, plants) imply complex regulatory patterns, so these unicellular eukaryotes ("multicellular prokaryotes") imply advances in the complexity of gene regulatory patterns over the solitary or pluricellular prokaryotes.

The timing of the evolution of eukaryotes remains uncertain, partly because the earliest eukaryotes closely resembled prokaryotes morphologically and distinguishing a fossil eukaryote from a fossil prokaryote among unicells is most difficult. Criteria aiding in the identification of early fossil eukaryotes include large size and the distinctive morphology (which implies multicellularity), but even such evidence is not perfectly foolproof. For example, millimeter-sized tubular, ribbonlike, and disc-shaped carbonaceous films interpreted as algae have been described from an early Proterozoic sequence (Hofmann and Chen, 1981). However, these fossils generally lack the details of cellular preservation and internal structure that are necessary for confident assignment to the eukaryotes (Horodyski and Bloeser, 1983). They may well be carbonaceous films produced by the compression and degradation of prokaryotic cellular aggregates.

Early fossils that do suggest the attainment of a eukaryote grade of organization are the acritarchs, which appear in clastic rocks around 1,400 Ma old (Horodyski, 1980; Vidal and Knoll, 1982). Acritarchs are organic-walled, acid-resistant, usually unicellular microfossils of uncertain taxonomic affinities (Downie, 1973). They are found abundantly preserved in clastic rocks, whereas most known microfossils more than 1,400 Ma in age are found preserved in silicified stromatolitic rocks. Acritarchs are regarded as part of the marine phytoplankton (Downie, 1973; Vidal, 1976). Middle and late Proterozoic acritarchs are significantly larger (greater than 60 μm) than typical prokaryotes and are generally interpreted as eukaryotes. Thus a possible interpretation of the evidence is that, since these earliest abundant eukaryotes were phytoplankters, the eukaryotic body plan or "cell plan" was developed with the invasion of the adaptive zone of marine phytoplankton.

We imagine that host cells possessing appropriate symbionts were part of a large benthic microbiota (though there is no fossil record of this) and that some of these proved preadapted to invasion of the planktonic realm. Members of any such benthic microbiota would have become suspended in the water column quite frequently as a result of wave activity or other turbulence; physical access to this habitat would certainly be wide open. Populations of microbes that could grow and reproduce in the water column facultatively could then simply remain there indefinitely and be subject to additional selection for stabilization of their ground plans at the eukaryotic level. Perhaps the cells most likely to become facultative plankters were those which lived on the surface of fine-grained substrates in shallow water. In such a habitat, exposure to solar radiation could favor the acquisition of photosynthetic capability as a ready source of supplemental energy for the cell. Such an adaptation would be roughly analogous to the employment of zooxanthellae by hermatypic corals and some reef mollusks (*Tridacna, Corculum*). Once these cells were in the plankton, the supplies of organic foods would have been much less concentrated than on the seafloor, and photosynthesis should have become obligatory. Indeed, some marine protozoans, including tropical benthic and planktonic foraminifera, harbor photosynthetic symbionts today (Loeblich and Tappan, 1964, 70).

The vast majority of modern marine eukaryotic phytoplankters possess "hard parts" of some sort; many have mineralized elements throughout most or all of their life cycles, others have tough organic walls at some life stage. We therefore infer that the acritarch wall structures were probably evolved with the invasion of the planktonic zone, to which they were adaptive. Thus we further suggest that the rise of a phytoplankton was associated with the origin of a cell plan that included preservable structures, leaving a fossil record to commemorate the event.

The status of the host ingestor cell, which plays such a major role in endosymbiont theories of eukaryote origins, is that of a hypothetical organism, much like the archetypes long hypothesized as ancestors to various animal phyla. No ingesting or phagocytizing prokaryotes are known (Stainer, 1970). The postulated host would have to have possessed several attributes atypical of prokaryotes. It should have lacked a cell wall, fed via ingestion of particles, thus gaining the potential to harbor endosymbionts, and possessed a genetic regulatory potential outstripping that of any prokaryote of which we now have knowledge. Perhaps the regulatory potential was invested in a nuclear organization unlike that of known prokaryotes. Whether the host was actually at the prokaryote grade in terms of genome structure and organization is moot.

Woese and Fox (1977) also postulate that the eukaryote grade was attained by a clade containing a unique host proto-eukaryote not found in any of the other extant prokaryotic alliances but descended from a common pre-prokaryotic ("progenote") ancestor. This scenario fits the facts well. The ingesting mode of life postulated for the host cells requires them to deal with the outside world more intimately than do noningestors, since they must discover and engulf foreign particles that must then be appropriately assimilated. Therefore, the host must have had a genetic regulatory capacity advanced for a prokaryote (or, if the host was not a prokaryote itself, different from prokaryotic regulatory systems) and was thus preadapted not only to capture but to amalgamate endosymbionts. There is evidence (Mahler et al., 1981) that the genome of mitochondria contains introns and exons, thus displaying a structure that has been thought to characterize the eukaryotes. Yet, ribosomal RNAs in mitochondria are allied to prokaryotic types, and indeed a purple nonsulfur bacterial ancestry for mitochondria is suggested by cytochrome sequencing (see Doolittle, 1980). Therefore, the genome of living prokaryotes may have been "streamlined" from more complexly structured ancestral conditions, of which the mitochondrial condition is a relic (Mahler et al., 1981).

In our scenario, the gap between prokaryotes and eukaryotes is created partly by the failure of ingesting prokaryotic clades to survive, partly by the unique organization of the eukaryotic genome that may be associated with the amalgamation of prokaryotic endosymbionts as organelles. If modern prokaryotes have indeed been streamlined insofar as their primitive genome structures are concerned, then their gene regulatory systems may have been simplified as well, which may have widened the gap still further.

Thus, we view the rise of complexity of organization from progenote to prokaryote to unicellular eukaryote to multicellular eukaryote grades as reflecting an evolutionary elaboration of genetic regulatory capacity, each major step also involving the invasion of a novel adaptive zone. The invading lineages arose from among organisms that were preadapted, fortuitously insofar as the subsequent invasion was concerned, to conditions in the new zone, a zone that happened to require or to be especially vulnerable to organisms of greater complexity. Postadaptive radiations then diversified the invaders to produce large successful clades at the higher level of complexity.

In summary, the sequence of adaptive steps leading to vascular plants should include the following (fig. 2.1). First, there was a rise from the progenote to the prokaryote grade. This step may have been associated with the development of an integrated cell with all the regulatory capacity that implies. At this grade, primitive life systems became freed from dependence upon prebiotic organic materials or biotic metabolites as energy sources. The next step would be the appearance among the lineages of prokaryotes of ingestor cells that preyed upon other prokaryotes and perhaps scavenged. These cells

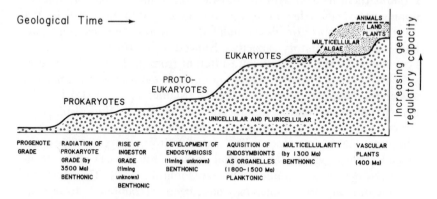

Fig. 2.1. Summary of postulated steps of rising sophistication in gene regulatory systems during the evolution of vascular plants from the earliest cells. The largest step is suggested to be related to the origin of the eukaryotic condition and to be associated with the invasion of the planktonic adaptive zone.

represented an adaptive zone distinct from autotrophic prokaryotes and may have possessed an unusual genome structure by the standards of today's prokaryotes. It is among such lineages that facultative endosymbiosis is postulated to have become established, creating an adaptive zone characterized by organisms with a supplemental energy source. Next, the establishment of endosymbiosis was associated with a further rise in the complexity of gene regulatory systems, and the coopting of the endosymbionts as organelles within a complex unicell involved a further sophistication and elaboration of gene regulatory functions. Endosymbiosis essentially created the eukaryotic ground plan for plantlike cells; we suspect that the pressures for consolidation of the symbionts into an integrated organism arose from the assumption of planktonic life. Multicellular algae then evolved several times from unicellular eukaryotes, which clearly possessed a genomic structure preadaptive to the production of genetic blueprints for multicellular body plans. Thus, multicellular algae may represent a return to a benthic habitat and perhaps originated as a benthic stage during the life cycle of essentially planktonic organisms that then seized an adaptive opportunity to become larger, more elaborate, and essentially benthic forms. Vascular plants arose from multicellular green algae (or perhaps directly from a unicellular ancestor; Stebbins and Hill, 1980). The versatility of the vascular plant genome is evidenced in the vast arrays of plant types that now clothe our green planet.

Relations to Major Environmental Events

It is common to attempt to account for important adaptive innovations as being responses to events in the history of the Earth's environment. However, factors in the origin of novel types are usually more obscure than are factors in their success, and the unicellular autotrophic eukaryotes are no exception. Although this chapter attempts to explain some factors in their origin, we cannot relate our adaptive scenario to any particular key environmental changes. On the other hand it is possible to identify some factors as requisite to their development and others as plausible if speculative contributors to their rise to dominance.

It has been speculated for some time that the presence of persistent free oxygen is a prerequisite for the attainment of the eukaryotic condition. This notion is supported by the fact that eukaryotes are mainly aerobic, except for those that are secondarily modified for an anaerobic existence. Furthermore, unlike the membranes of most prokaryotes, eukaryote membranes contain steroids, which can only be synthesized in the presence of oxygen (see Margulis, 1981). Free oxygen was first available about 2,000 Ma ago (Cloud, 1976). This leaves a gap on the order of 600 Ma between the appearance of persistent free oxygen and the inferred appearance of eukaryotic phytoplankters. Because the rate at which oxygen levels increased during this interval is uncertain, it is possible that oxygen remained at such a low concentration that it provided a check on the development of eukaryotes. On the other hand it is quite possible that the eukaryotes employed oxygen simply because, when they developed, it was there in some abundance.

An environmental change affecting the success of the eukaryotic phytoplankton may have been associated with changes in tectonic style during the middle Proterozoic. The earlier tectonics appear to have been different from younger styles (Ronov, Khain, and Seslavinskiy, 1980; Goodwin, 1981). Although some geologists have suggested that the early Proterozoic (beginning some 2,500 Ma ago) signaled the beginning of Phanerozoic macroplate tectonics (Hoffman, 1980), the majority of early to middle Proterozoic mobile belts exhibit features that indicate limited ocean opening, no consumption of ocean floor and a low degree of continentality (Kröner, 1981). This inferred tectonic style of intercratonic troughs, intermediate between the Archean greenstone belt-microplate phase and Phanerozoic macroplate-plate tectonics (Goodwin, 1981), may have had some effect on the early evolution of phytoplankton. With little seafloor spreading there would have been a low input of mineral nutrients from weathered oceanic crust; continentally derived abiotic nutrients would have been the major source. Consequently, the open ocean may have been starved. About 1,200 Ma ago, widespread rock assemblages characteristic of passive and active continental margins first occur, marking the onset of the Phanerozoic style of global tectonics (Kröner, 1981). Active continental margins, seafloor spreading, and greater continentality should have provided a new type of nutrient dynamics in the seas and perhaps provided for the first time an opportunity for the widespread establishment of phytoplankton communities in the oceanic realm.

Our inability to associate the origins of the many new grades and adaptive novelties we have discussed with key physical environmental events is symptomatic of the general situation in macroevolutionary studies. To us, this strongly suggests that novelties are not created by deterministic forcing factors that act to fill immediately any open nooks or crannies of the biosphere. Rather, adaptive opportunities may lie fallow for many tens of millions of years until the boundaries of the adaptive zones involved are broached through a fortuitous concomitance of circumstance and a new adaptive type is developed.

Summary

The morphological variety and complexity achieved by multicellular eukaryotic kingdoms such as the Plantae is underpinned by the evolution of gene regulation patterns. It is, therefore, hypothesized that the ability of eukaryotes to produce such complexity depends upon their possession of sophisticated gene regulatory systems and that the lack of truly complex ground plans among prokaryotes reflects the relatively simple organization of their gene regulation mechanisms. We hypothesize that the origin of eukaryotes, associated with the amalgamation of prokaryotic symbionts into a host ingestor, required novel regulatory mechanisms to effect integration of what became effectively a "multicellular prokaryote" — the eukaryotic cell.

Acknowledgments

We thank Andrew H. Knoll, Harvard University; Bruce Tiffney, Yale University; Carmen Sapienza, University of Utah; and Robert Trench, UCSB, for constructive comments. Research on which this paper is based was supported in part by NSF Grant EAR82–05809 (to SMA) and by NASA Grant NAG 2–73 (to JWV). This is contribution no. 120 of the Preston Cloud Research Laboratory, Department of Geological Sciences, University of California, Santa Barbara.

References

Bogorad, L. 1975. Evolution of organelles and eukaryotic genomes. *Science* 188:891–98.

Britten, R. J., and E. H. Davidson. 1969. Gene regulation for higher cells: A theory. *Science* 165:349–57.

———. 1971. Repetitive and non-repetitive DNA sequences and a speculation on the origins of evolutionary novelty. *Q. Rev. Biol.* 46:111–33.

Campbell, J. H. 1982. Autonomy in evolution. In *Perspectives on evolution*, ed. R. Milkman, 190–201. Sunderland, MA: Sinauer Association.

Carlile, M. 1982. Prokaryotes and eukaryotes: Strategies and successes. *Trends Biochem. Sci.* 7:128–30.

Cavalier-Smith, T. 1981. The origin and early evolution of the eukaryotic cell. *Symposium of the Society for General Microbiology* 32:33–84.

Cloud, P. 1976. Beginnings of biospheric evolution and their biogeochemical consequences. *Paleobiology* 2:351–87.

Dobzhansky, Th., F. J. Ayala, G. L. Stebbins, and J. W. Valentine. 1977. *Evolution*. San Francisco: Freeman and Co.

Doolittle, W. F. 1980. Revolutionary concepts in evolutionary biology. *Trends Biochem. Sci. Ref. Ed.* 5:146–49.

Downie, C. 1973. Observations on the nature of acritarchs. *Palaeontology* 16:239–59.

Geitler, L. 1932. Cyanophyceae. In *Dr. L. Rabenhort's Kryptogamenflora von Deutschland, Österreich und der Schweiz*, ed. R. Kolkwitz, vol. 14. Leipzig: Akademische Verlagsgesellschaft; New York: Johnson Reprint Corp., 1971.

Goodwin, Am. M. 1981. Precambrian perspectives. *Science* 213:55–61.

Hoffman, P. F. 1980. Wopmay Orogen: A Wilson cycle of early Proterozoic age in the northwest of the Canadian shield. In *The continental crust and its mineral deposits*, ed. D. W. Strangway. *Spec. Pap. — Geol. Assoc. Can.* 20:523–49.

Hofmann, H. J., and Chen Jinbiao. 1981. Carbonaceous megafossils from the Precambrian (1800 Ma) near Jixian, northern China. *Can. J. Earth Sci.* 18:443–47.

Horodyski, R. J. 1980. Middle Proterozoic shale-facies microbiota from the Lower Belt Supergroup, Little Belt Mountains, Montana. *J. Paleontol.* 54:649–63.

Horodyski, R. J., and B. Bloeser. 1983. Possible eukaryotic algal filaments from the Late Proterozoic Chuar Group, Grand Canyon, Arizona. *J. Paleontol.* 57:321–26.

Hunkapiller, T., H. Huang, L. Hood, and J. H. Campbell. 1982. The impact of modern genetics on evolutionary theory. In *Perspectives on evolution*, ed. R. Milkman, 164–89. Sunderland, MA: Sinauer Association.

Kröner, A. 1981. Precambrian plate tectonics. In *Precambrian plate tectonics*, ed. A. Kröner, 57–90. Amsterdam: Elsevier.

Loeblich, A. R., and H. Tappan. 1964. *Treatise on invertebrate paleontology: Protista 2*, vol. 1. Lawrence, KS: Geological Society of America and Univ. of Kansas Press.

Mahler, H. R., P. S. Perlman, D. K. Hanson, and S. Dhawale. 1981. Introns in mitochondria and their possible significance in evolution. In *Evolution today*, ed. G. G. E. Scudder and J. L. Reveal, 245–56. Pittsburgh: Hunt Institute for Botanical Documentation, Carnegie-Mellon University.

Margulis, L. 1970. *Origin of eukaryotic cells*. New Haven: Yale Univ. Press.

———. 1981. Symbiosis in cell evolution. San Francisco: W. H. Freeman.

Mereschkovsky, C. 1905. Über Natur und Ursprung der Chromatophoren im Pflanzenreiche. *Biol. Zentralbl.* 25:593–604.

Parish, J. H. 1979. Myxobacteria. In *Developmental biology of prokaryotes*, ed. J. H. Parish, 227–53. Berkeley: Univ. of California Press.

Raff, R. A., and H. R. Mahler. 1975. The symbiont that never was: An inquiry into the evolutionary origin of the mitochondrion. *Symposium of the Society for Experimental Biology* 29:41–92.

Ronov, A. B., V. Ye. Khain, and K. B. Seslavinskiy. 1980. Lower and middle Riphean lithologic complexes of the world. *Int. Geol. Rev.* 24:509–23.

Schimper, A. F. W. 1883. Über die Entwicklung der Chlorophyllkörner und Farbkörper (I. Teil). *Botanische Zeitung* 41:105–14.

Schopf. J. W., and D. Z. Oehler. 1976. How old are the eukaryotes? *Science* 193:47–49.

Stainer, R. Y. 1970. Some aspects of the biology of cells and their possible evolutionary significance. *Symposium of the Society for General Microbiology* 20:1–38.

Stainer, R. Y., and C. B. von Niel. 1962. The concept of a bacterium. *Arch. Mikrobiol.* 42:17–35.

Starr, M. P., and J. M. Schmidt. 1981. Prokaryote diversity. In *The prokaryotes*, ed. M. P. Starr, H. Stolp, H. G. Trüper, A. Balows, and H. G. Schlegel, 3–42. Berlin: Springer-Verlag.

Stebbins, G. L., and G. J. C. Hill. 1980. Did multicellular plants invade the land? *Am. Nat.* 115:342–53.

Taylor, F. J. R. 1974. Implications and extensions of the serial endosymbiosis theory of the origin of eukaryotes. *Taxon* 23:229–58.

Trench, R. K. 1982. Physiology, biochemistry, and ultrastructure of cyanellae. *Prog. Phycol. Res.* 1:257–88.

Valentine, J. W., and C. A. Campbell. 1975. Genetic regulation and the fossil record. *Am. Sci.* 63:673–80.

Vidal, G. 1976. Late Precambrian microfossils from the Visingsö beds in southern Sweden. *Fossils Strata* 9, no. 7.

Vidal, G., and A. H. Knoll. 1982. Radiations and extinctions of plankton in the late Proterozoic and early Cambrian. *Nature* 297:57–60.

Woese, C. R., and G. E. Fox. 1977. The concept of cellular evolution. *J. Mol. Evol.* 10:1–6.

DAVID J. CHAPMAN

Geological Factors and Biochemical Aspects of the Origin of Land Plants

Introduction

In this chapter I will consider a possible relationship between the land migration and (1) the development of an ultraviolet (UV) radiation screen and (2) the development of an atmosphere with enough oxygen to produce this screen and to provide for the biosynthesis of cutin, the plant waterproofing agent. Both developments may be correlated with an increasing partial pressure of oxygen. The discussion will be extended to a consideration of lignin biosynthesis vis-à-vis O_2 partial pressures and to the possible connection with the explosive radiation of land plants subsequent to the migration. The UV screen concept has already been proposed (Lowry, Lee, and Hebant, 1980), but in a more qualitative sense; I attempt in this chapter to provide a quantitative evaluation.

Characteristics of Land Plants

It is necessary to consider exactly what is meant by the term *land plant*. A strict definition will not be given; instead, a description based on functional considerations will be provided so that the reader will know what groups of plants are encompassed by this term.

The essential difference between aquatic and terrestrial plants is their handling of water relations. Aquatic plants inhabit a water-sufficient environment (those living in hypersaline areas are regarded as specialized derivatives of aquatic plants). Terrestrial plants, in contrast,

23

live in a water-deficient environment. Their adaptation to this problem of water deficiency is a key feature of hypotheses concerning their migration to the land. This statement implies that the ancestors of the land flora were primarily, or originally, aquatic, and it is accepted that these ancestors were the aquatic green algae.

Fungi in nonaquatic environments also encounter problems of water deficiency and have acquired a number of adaptations to deal with them. As heterotrophs and chemotrophs, however, they use organic carbon as their source of cellular carbon and energy and hence almost certainly appeared on the terrestrial scene after the appearance of a land flora and/or fauna.

In this discussion the term *land plant* will refer to plants permanently adapted to a terrestrial or water-deficient habitat. This is a functional description, designed for this paper. Aquatic plants such as *Myriophyllum*, *Elodea*, *Sagittaria*, and *Potamogeton* are secondarily aquatic, having "migrated" back to water and adapted with a secondary reduction of lignified tissue, formation of aerenchyma, and the acquisition of an algalike morphology. Many bryophytes and algae also inhabit damp environments and have acquired some but not all of the adaptations needed for a terrestrial existence; they are restricted to a semiaquatic or semiterrestrial habitat. There are additionally a number of drought-resistant bryophytes and terrestrial (for example, soil) algae.

A plant, to be considered permanently adapted to a terrestrial habitat (Lewis, 1980; Miller, 1970; Stebbins and Hill, 1980; Walker, 1983), must possess the following characteristics.

1.   A soil-bound absorbing region (roots or rhizoids).
2.   Waterproofing on the aerial parts to keep water in and "toxic" solutions out (Walker, 1983). Waterproofing is provided by the polymer cutin.
3.   A water transport system. Water transport is principally the function of the xylem tissue and its associated lignin, which serve also as mechanical support for the aerial part of the plant. In the case of some mosses, however, water transport and conduction is provided by hydroids and leptoids, which are nonlignified tissues.

The *transitional* plant, which can be accepted as semiaquatic, but not permanently adapted to land, would have to meet the first two requirements. A permanently adapted plant (that is, a land plant) would be expected to meet all three.

In this context it should be noted that the term *permanently adapted* does not mean "restricted" but rather "capable of survival of growth."

With these descriptions and caveats in mind, one can include in land plants both vascular plants (such as *Myriophyllum*, *Elodea*, *Sagittaria*, and *Potamogeton*, mentioned above) and bryophytes. Aquatic algae, obviously, do not need to meet these requirements. So-called terrestrial algae circumvent many water-deficit problems with a tolerance of desiccation or a mucilaginous sheath to minimize water loss. Growth of such terrestrial algae is usually very slow.

Other nonessential or secondary adaptations (table 3.1) would

Table 3.1 Adaptation of Algae to a Terrestrial Existence and the Needed Modifications

Green Algae to Semiaquatic Existence
    Development of waterproofing: cutin
    Development of pores for gas exchange: simple pores
    Development of UV shield (in absence of sufficient O_3 shield)
    Development of desiccation-resistant spores: sporopollenin
    Partial substrate-attached absorbing region

Semiaquatic to Totally Terrestrial Existence
    Development of water-conducting tissue: lignin
    Mechanical support for aerial plants: lignin
    Soil-bound absorbing region (roots or rhizoids)
    Mycorrhizal system with roots: fungal association
    Reproduction independent of swimming gametes or zoospores
    Reduction of surface-to-volume ratio
    Improved gas exchange (stomata)

include (Jeffrey, 1962; Stebbins and Hill, 1980) the development of a reproductive mechanism independent of swimming gametes or zoospores; a zygotic resting stage on the aerial structure; thick-walled spores and a tendency toward the reduction of the surface-to-volume ratio; and an efficient gas-exchange mechanism. None of these may be considered essential to the migration. They are later developments that contributed to the subsequent adaptive radiation of the land flora. Another important adaption and one that has received little attention is the development of mycotrophism or mycorrhizal associations (Malloch, Pirozynski, and Raven, 1980; Pirozynski and Malloch, 1975). This system, describing terrestrial plants (Pirozynski and Malloch, 1975, 153) as the product of a "continuing symbiosis of a semiaquatic ancestral green alga and an aquatic fungus — an öomycete," is important to the solution of nutrient starvation and is argued to be an important factor in the development of a land flora.

Land Plants and Presumptive Green Algal Ancestors

Seventy-five years ago Bower's (1908) classic work, *The Origins of a Land Flora*, was published. In the intervening years two recurrent but essentially unanswered questions have persisted as the sources of considerable speculation: What was the putative algal ancestor of embryophytes and what was the nature of the first land plants?

The postulated algal ancestor was sought usually in the Chaetophorales (for example, *Fritschiella*) — one of the primary reasons for this hypothesis being the heterotrichous habit found in this order, which provided a gross morphological similarity to what many have envisaged as the structure of the primitive plants. At first glance this seems to be a perfectly reasonable assumption. Certainly at the level of gross morphology there are no other candidates that can be considered as "more obvious" ancestors. However, such arguments are prejudiced by the sparse fossil record of identifiable early land plants, views about the phylogenetic position of the Bryophyta, and the morphology of simple extant vascular plants. By any account there is a vast morphological chasm between any putative algal ancestor and any as yet identified primitive land plant, fossil or living.

The last fifteen years have seen a total reevaluation of green algal taxonomy and phylogeny (Melkonian, 1980, 1982; Moestrup, 1978; Pickett-Heaps, 1975; Stewart and Mattox, 1975, 1978). The impetus for this change derives from ultrastructural studies, especially on mitotic and microtubular systems and flagellar root structures, with support from biochemical data. The proposed phylogenetic model is shown in figure 3.1, and a chlorophytan taxonomy is given in table 3.2. When comparisons with higher plants are then made it becomes obvious that

Table 3.2   Taxonomy of the Chlorophyta

Class:	Ulvaphyceae	
	Orders:	Dasycladales
		Codiales
Class:	Chlorophyceae	
	Genera:	*Fritschiella*
		Stigeoclonium
Class:	Charophyceae	
	Orders:	Charales (*Chara*, *Nitella*)
		Conjugales (*Spirogyra*)
		Klebsormidiales
		Coleochaetales (*Coleochaete*)

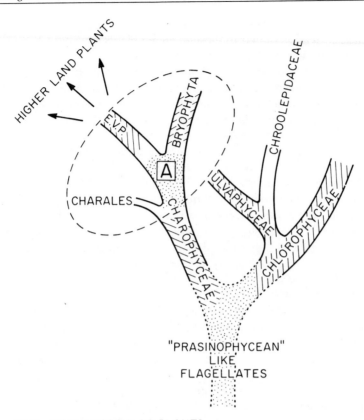

E.V.P. = EARLY VASCULAR PLANTS

Fig. 3.1. Phylogenetic scheme of the "green plants." Section inside the ellipse represents the position of the land migration discussed in the text. *A* represents the evolutionary position of the hypothetical migratory green alga. The various taxa and groups of organisms are delineated by shaded and stippled areas.

the old Chaetophorales must be abandoned as the algal ancestral stock and that more likely candidates are the Klebsormidiales and Coleochaetales, now in the class Charophyceae (*sensu* Stewart and Mattox, 1975, 1978). Included in the "abandoned" order were such genera as *Fritschiella* and *Stigeoclonium*, both of which have a basal and aerial filamentous part to their thalli. The grounds for this rejection are the major differences between these algae and higher plants in flagellar base and root structure (Melkonian, 1982; Stewart and Mattox, 1975, 1978) and mitotic-cytokinetic behavior (Pickett-Heaps, 1975, 1979).

Chaetophoraleans with mitotic spindle, cell division, and flagellar root and base structures similar to those of higher plants have been "transferred" to the Charophyceae, which includes *Chara* and *Nitella* (order Charales). Superficially, this is an unhappy state of affairs. The morphological gap between a primitive presumptive land plant and the green algal stock is now greater. A "best" green algal ancestor now becomes *Coleochaete*, a simple heterotrichous form with a prostrate disc and erect hairs. It would apparently be simpler to formulate hypotheses regarding the ancestors of the vascular land plants if (1) there were a charophycean alga with distinct tissues, an erect thallus, and some form of identifiable (preferably terminal) sporangia or (2) if there were a discernible semiaquatic fossil with a more algalike morphology than *Cooksonia*, a very early fossil vascular land plant. The advanced members of the Charophyceae (Charales) do not meet this requirement and indeed one can argue that they are an evolutionary side branch of the charophyceae, not closely related to the line giving rise to the vascular land plants.

The argument whether unicellular or multicellular green algae were the transitional stock (Stebbins and Hill, 1980) does not affect the discussion pertinent to vascular land plants. It is important in considering nonvascular land plants, and these authors have proposed that the main evolution of the Charophyceae occurred on moist surfaces on land. The extant Charophyceae, which are almost entirely freshwater, would thus be secondarily aquatic. The recovery of *Cooksonia*, from marine sediments in some instances, is not proof of an aquatic habitat.

The Fossil Record and the Time Frame and of the Land Migration

The first land plant, or at least one definitely identifiable as *vascular* and terrestrial, is generally accepted (Banks 1975a, 1975b; Taylor, 1982) to be similar to *Cooksonia*. Other primitive plant remains with ligninlike chemistry, but not necessarily vascularized, include *Eohostimella* and *Sciadophyton* (cf. Niklas, 1980); primitive and apparently nonvascularized remains include *Horneophyton, Steganotheca*, and *Spongiophyton* (Niklas, 1980; Taylor, 1982). The nonvascular nature of the two former may be doubtful, and proof of a green algal ancestry for *Spongiophyton* is lacking. Indeed, green algal affinities of many nonvascular remains are questionable; consider, for example, the paleobiochemical clustering of such fossils as *Prototaxites, Protosalvinia*, and *Nematothallus* with brown and red algae (Niklas, 1980).

Plant remains (for example, *Eohostimella*) with ligninlike biochemistry (vascularized?) have been dated to the lower Silurian (Niklas,

1980; Niklas and Pratt, 1980), 475 million years ago (cf. also Schopf, 1970). *Cooksonia*, the most commonly accepted first vascular land plant, dates back to the upper Silurian, 405 million years ago. The reader is referred to Edwards, Bassett, and Rogerson (1979) for a discussion of the earliest *Cooksonia*.

Nonvascularized remains have been recorded from the lower (Pratt, Phillips, and Dennison, 1978; Niklas and Smocovitis, 1983; Strother and Traverse, 1979) and upper Silurian. Perhaps the most notable of these is *Parka*, which can be regarded as showing morphological (Niklas, 1976b) and paleobiochemical similarities (Niklas, 1976a) with hypothetical ancestral green algae. Fossil green algae have been identified in the late Cambrian-Ordovician (Wray, 1977), 550–500 million years ago. These forms are advanced Dasycladales and Codiales (Herak, Kochansky-Devide, and Gusic, 1977), two orders that are not in the general evolutionary line to the higher plants. The ancestral stock (the charophytes) has representatives dating to the upper Silurian (Grambast, 1974). Assumed green algae are generally considered to be much older, but it is difficult to establish their true affinities vis-à-vis the Charophyceae, Chlorophyceae, or Ulvaphyceae. The net result of this fossil record is to suggest (Taylor, 1982, 171) that "land inhabiting plants were well established by the beginning of the Silurian, and possibly much earlier in the Ordovician." It is important to remember that the dates of the fossil record represent only the first *identification*, not the first actual occurrence, which may well have been earlier. In the remainder of this paper I will attempt to show that these data, and those for vascularization (dated to the late Silurian, 415 million years ago), both established from the fossil record, agree with the dates developed from biochemical/environmental arguments, which are based upon the premises, outlined in the introduction concerning the role of the ultraviolet screen and an oxygenated atmosphere needed to produce this screen and to support certain biochemistries essential for the establishment of a land flora.

Biochemical Requirements and the Environment

The arguments outlined above suggest that the migration to the land and the adaptation to the terrestrial habitat required two major *biochemical* innovations not required by (or found in) the presumptive green algal aquatic ancestor: (1) development of cutin and (2) elaboration of phenylpropanoid metabolism. The first meets the requirement for a waterproofing agent. Phenylpropanoid metabolism could have provided a solution to the potential need for an ultraviolet screen.

Modeling of atmospheric changes over geological time (see below) suggests that the ozone screen may have been insufficiently developed at the time of the transition (perhaps 600 million years ago) to provide an "external" UV screen for the emerging land flora. If this were the case, then one can argue that the transition would have been dependent in part upon, or facilitated by, the development of an "internal" (plant-originated) UV screen. This assumption has been incorporated into the material of this paper. Although lignin is not essential for the transition, its subsequent appearance, which provided for the morphological elaboration of erect structures and production of water-conducting xylem tissue, presumably soon after the transition, was important in the elaboration of the land flora. The reader is referred to two recent publications (Niklas, 1976c, 1982) for a discussion of possible additional roles for these and other secondary metabolites. The question remains, were the environmental changes that may have been occurring at the time of the transition instrumental in facilitating these innovations?

Cutin: Waterproofing and Protection

Waterproofing and protection against external toxic solutes is the raison d'être for cutin. In aquatic systems waterproofing is not a real problem, and toxic substances can be expected to be diluted. Protective coatings can also be provided by heavy mucilaginous layers, thick cellulose walls, or sporopollenin. However, cutin appears to be the best waterproofing agent. Besides its excellent water-repellent properties, it lacks the rigidity of sporopollenin but, unlike mucilage, it is rigid enough to allow pores (stomata) to open and close for critical gas exchange.

Cutin is composed of a phenylpropanoid, C_{16} (or C_{18}) polyoxy fatty acid polymer (Kolattukudy, 1980). The structure of cutin reveals a polymer whose monomeric unit is coumaric acid complexed to a number of mono- and dihydroxy C_{16} or C_{18} fatty acids. The hydroxylation-epoxidation of the fatty acid requires O_2. The synthesis of the hydroxy-palmitate (C_{16}) moiety involves two hydroxylations (NADPH/O_2) and that of the trihydroxystearate (C_{18}) involves three similar oxygenation steps. Cutin is a highly oxygenated substance, as are the phenylpropanoids, and is essential to land-plant development.

The migratory alga, or the very first semiaquatics, would have had to develop the capacity to synthesize cutin very early. Besides acting as a waterproofing agent, cutin can also be regarded as playing a protective role (Walker, 1983) — keeping toxic or harmful solutions out. No algae have been identified as synthesizing cutin (they really have no need for it), whereas the semiaquatic or terrestrial mosses and liverworts (Kolattukudy, 1980, 1981) in the Bryophyta do produce the substance. The retention of cutin in aquatic angiosperms is obviously a residual trait retained during the recolonization of the aquatic environment.

There is currently no information about the partial pressure of O_2 needed for cutin biosynthesis, either for the total compound or the individual coumaric acid and hydroxy fatty acids. What information is available (Chapman and Schopf, 1983) suggests that there is generally a critical level of O_2 partial pressure for an individual enzymatic reaction (called systemic biochemistry) and another critical level for the maintenance of organismic or physiological functions, or a complete biosynthetic sequence of serial enzymatic steps. The O_2 concentration necessary for half-maximal activity of individual aerobic (O_2-requiring) enzymes that have been examined (see citations in Chapman and Schopf, 1983) is 0.002 PAL O_2 (the PAL, or present atmospheric level, of O_2 is 0.2 atmospheres).

The half-maximal O_2 partial pressure necessary to produce a full cellular complement (that is, concentrations normally encountered) of oxygenated compounds requiring oxygen in biosynthesis is 4×10^{-4} atmospheres ($= 0.002$ PAL O_2). Physiological functions (growth, respiration, mitosis) show half-maximal activity (or inhibition, in the case of mitosis) at about 0.02 PAL O_2 — an order of magnitude higher. The oxygen level at which amphiaerobic organisms (Chapman and Schopf, 1983) switch from an anaerobic to an aerobic metabolism (Pasteur point) is approximately 0.01 PAL O_2. These values suggest that although individual biosynthetic steps in coumaric acid synthesis and fatty acid hydroxylation may have been operative at lower O_2 partial pressures, aggregate pathways involving a number of O_2-requiring steps might not have occurred until the level was in the range of 0.02 PAL O_2. With the exception of certain carotenoids in the choloroplasts, biosynthetic sequences involving numerous O_2-requiring steps are not common in green algae. It is in the land plants that one usually finds the elaboration of polyoxygen metabolites, in which the O_2-requiring steps frequently occur as "terminal add-ons" to preexisting pathways.

Let us assume that an O_2 level of 0.02 PAL is needed for cutin biosynthesis. At what time in the past was this level reached? Estimates of oxygen levels are based on geological and paleobiological data and on atmospheric modeling. These estimates are usually given only as limits (maximum or minimum, earliest or latest) within which one can work. Walker et al. (1983) have argued that it is unlikely that an oxygen level of 0.1 PAL was reached prior to 1.7 Ga ago and that in fact a stable aerobic environment was established apparently no earlier than 1.7 Ga ago. These are *not* the dates of the attainment of this O_2 level but only the earliest possible date. Rhoads and Morse (1971) have examined the metazoan fossil record in the light of modern metazoan distribution in low-O_2 environments. They have suggested that an O_2 level of 0.01 PAL was reached about 0.7–0.8 Ga ago and a minimum of 0.1 PAL was reached at the lower Cambrian boundary of 570 million years ago. These sets of dates provide a window, admittedly broad, within which the migration could have occurred. The dates provided by Rhoads and Morse (1971) coincide quite well with the fossil record which indicates the presence of land flora approximately 500–450 million years ago and a presumably earlier date for the transition to the land.

Phenylpropanoids and UV Screening

A land migration could not have occurred until either sufficient ozone existed to screen out the UV irradiation or the antecedents to the migratory flora had developed their own internal UV screen. Although O_2-O_3 calculations (see below) would suggest that an internally produced (in vivo) UV screen was not necessary, such a screen may well have facilitated the migration if the O_3 levels at the time were marginal or had just attained the appropriate level. Several compounds typical of C_6-C_3 metabolism have good UV-absorbing properties, and it has previously been suggested (Lowry, Lee, and Hebant, 1980) that phenylpropanoid metabolism did in fact represent an early protective adaptation. The simplest UV-absorbing compounds are coumaric acid and its immediate derivatives. We can assume coumaric acid, as an intermediate to the ubiquinones, was formed in all organisms. Formation of its hydroxy derivatives in the land-plant antecedents has not been demonstrated. These derivatives have some potential as UV screens, but there are some unanswered questions.

Examples of compounds that screen UV light include the following.

	Absorption Spectrum λ max in nm (methanol)
Coniferyl alcohol	265
Dihydroxy dimethoxy stilbene	306–333
Coumaric acid	224–299
Ferulic acid	235–320
Sinapic acid	240–320

These substances are derivatives of phenylalanine and their formations are governed by a key enzyme, phenylalanine ammonia lyase. This enzyme has been detected in the chlorophycean *Dunaliella* and its presence may be assumed by the synthesis of ubiquinone in algae. We do not know to what extent the simple coumaric acid derivatives are stored free in the cell, as opposed to serving as intermediary metabolites for lignins, and dilignols of bryophytes. It may be that they themselves are insignificant in UV protection and that the important compounds would have been the flavonoids. The presence of flavonoids (Markham and Porter, 1969; Swain, 1980), elaborated from C_6-C_3 phenylpropanoids, has been observed in *Nitella* and *Chara* (Charales). Flavonoids have not been reported in any other green algae, so their presence in *Nitella* and *Chara* suggests that flavonoid production may be a feature of the advanced members of the Charophyceae, the group that presumably gave rise to higher plants. Considerably more comparative biochemical investigation is needed to offset the caveat that this distribution may be simply a reflection of the extent of the search. There is good documentation that bryophytes have the capacity to elaborate simple phenylpropanoids. This is shown by the capacity of bryophytes to synthesize lunularic acid (Gorham, 1977a, 1977b; Pryce, 1972) and a wide selection of flavonoids (Markham and Porter, 1978). It would be interesting to look for lunularic acid in the Charophyceae line. The algae examined by Gorham (1977a; cf. Pryce, 1972) are members of the Chlorophyceae and Ulvaphyceae, not on the line to higher plants. Figure 3.2 illustrates the basic pathways of phenylpropanoid metabolism, and figure 3.3 shows the structures of lunularic acid and two representative flavonoids found in bryophytes and *Nitella*. Many flavonoids, including those in *Nitella*, have UV-absorbing properties that could well enable them to serve as effective UV screens, provided there was an appropriate cellular location and content, a feature that is not known for either *Nitella* or the bryophytes.

UV-Absorption Spectra Properties of Nitella *Flavonoids*

	λ *max in nm (methanol)*
Lucenin	256–270
Orientin	255–267
Vicenin	273–333
Vitexin	270–336

These data from comparative biochemistry, admittedly sparse, do suggest that the putative ancestors of the migratory flora were capable of synthesizing their own UV screen, in the form of flavonoids, and that these compounds were probably present in the earliest land plants.

Two questions must now be resolved concerning the timing of the

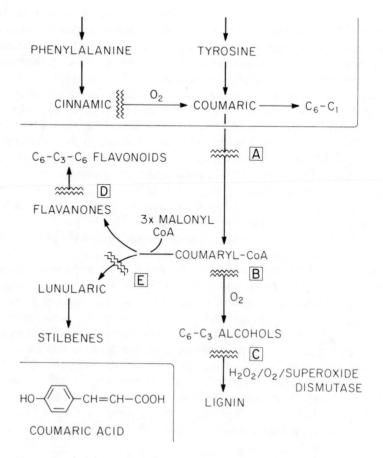

Fig. 3.2. Phenylpropanoid metabolism. Area inside the line at the top
represents metabolism common to all groups.

A. Coumaryl ligase. Appearance of this enzyme enabled the
biosynthesis of flavanones and stilbenes and the establishment of
an initial internal ultraviolet screen.

B. Monophenol oxygenase. This enzyme is necessary for the
elaboration of the C_6-C_3 alcohols, an O_2-requiring step. The
alcohols are precursors of lignin and the dimers of bryophytes.

C. Elaboration to the true lignins of higher plants representing a
complex of steps requiring superoxide, H_2O_2, and peroxidase.

D. Flavanone elaboration to the family of flavones.

E. Extension of A, which allowed synthesis of stilbenes in bryophytes.
This is a presumptive biosynthetic pathway.

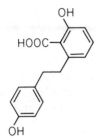

LUNULARIC ACID

In BRYOPHYTA: HEPATICOPSIDA

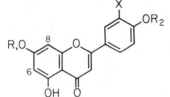

FLAVANONE SKELETON

In: CHAROPHYCEAE
BRYOPHYTA
(BRYOPSIDA)
(HEPATICOPSIDA)

Apigenin: $R_1 = R_2 = X = H$
   Vicenin: C-6, C-8 Diglycoside
   Vitexin: C-8 Monoglycoside

Luteolin: $R_1 = R_2 = H$; X = OH
   Lucenin: C-6, C-8 Diglycoside
   Orientin: C-8 Monoglycoside

Fig. 3.3. An example of the stilbenes (lunularic acid) and skeletal structure of the flavanones, which are synthesized by the Bryophyta and Charophyceae. These compounds may have represented early internal, or cellular, screens to UV radiation. Synthesis of these compounds is indicated by step *A* in figure 3.2.

invasion of the land. First, was a sufficient ozone screen present at that time, and if not, was an internal UV screen present in the earliest land plants? Second, what is the time frame of the appearance of these two features? Calculations of early ozone levels depend upon theoretical calculations of modern ozone production and upon the estimation of early oxygen levels, which are based upon calculations and estimates from the geological record. Walker et al., (1983) have pointed out the disparity in the various estimates but have suggested that the calculations of Kasting and Donahue (1980), who reevaluated the earlier calculations of other workers, are the most reliable. Kasting and Donahue found that an effective UV screen does not occur until an O_2 level of 0.1 PAL is reached. In the absence of a cellular UV screen (for example, flavonoids or other phenylpropanoids), a land migration is unlikely to have occurred prior to a time at which this level was reached. Walker et al. (1983) have indicated that it is unlikely that a level of 0.1 PAL O_2 was reached before 1.7 Ga ago. However, as mentioned earlier in the discussion of cutin, this represents an "earliest possible" date. In contrast, Rhoads and Morse (1971) have suggested from paleobiological considerations that a minimum of 0.1 PAL was reached at the lower Cambrian boundary of 570 million years ago.

Turning again to the option of an internal UV screen, we must still consider the time frame. There is an interesting catch to the biosynthesis of phenylpropanoids that bears on this evolutionary development. The biosynthesis of the polyhydroxy C_6-C_3 phenylpropanoids requires oxygen (fig. 3.2). Even though ammonia lyase activity may well have been operative in early green algae (*viz.*, phenylalanine ammonia lyase activity in *Dunaliella*; Loeffelhardt, Ludwig, and Kindl, 1973), elaboration of the aromatic amino acid phenylalanine to its polyhydroxy derivatives presumably would have been influenced by O_2 partial pressures. The same arguments for a minimal O_2 partial pressure that were used for cutin biosynthesis (see above) can be applied here. In other words, although individual steps in the biosynthesis of the polyhydroxy phenylpropanoids may operate at low O_2 levels (0.002 PAL), aggregate pathways to a polyhydroxy compound (for example, flavonoid) may require higher levels of about 0.02 PAL O_2.

An internal screen (for example, the presence of phenylpropanoids) is not universal among simple plants. It is frequently absent in fungi, and many green algae (those not on the line to the higher plants; that is, Chlorophyceae, Ulvaphyceae) may lack such a screen. Does this favor the option that an efficient O_3 screen was already present prior to the migration? Or, alternatively, does it suggest that there may have been later, continued migrations after the attainment of an O_3 screen? At this time it is probably impossible to distinguish between these two alternatives. For example, we know very little about mycorrhizal fungi and their capacity to synthesize potential internal chemical UV screens or their time of migration. It must also be remembered that the land migration may not have been a "one-event" occurrence and that a number of migratory "events" may have occurred while the O_3 shield was increasing; some organisms may have been developing an internal shield and some not.

Lignin: Water Transport and Support

Lignin, which is a polymer of C_6-C_3 phenylpropanoid compounds, is the least essential of the three requisites under consideration for land-plant colonization. Lignin is not required for the migration and the development of land plants with aerial parts. The presence of lignin surely contributed to the full evolution of the land flora by providing structural rigidity and the lignified vascular tissue. There is no evidence to suggest that the transitional plants could synthesize lignin, although they almost certainly could synthesize its precursors. Advanced green algae of the Charophyceae had acquired the capacity to synthesize C_6-C_3 phenylpropanoid precursors, as shown by the synthesis of flavonoids (Lowry, Lee, and Hebant, 1980; Swain, 1980) in *Nitella* and *Chara*. Green algae of the Chlorophyceae and Ulvaphyceae, which are not on the line to higher plants, apparently have not acquired the capacity to elaborate C_6-C_3 metabolism beyond coumaric acid (see fig. 3.2), as shown by the apparent failure to identify later biosynthetic compounds in these groups. Bryophytes can synthesize lunularic acid (Gorham, 1977a, 1977b; Pryce, 1972) and have, additionally, acquired the ability to modify the C_6-C_3 moiety and dimerize it (Miksche and Yasuda, 1978). These dilignols, however, are *not* true lignins.

Lignin formation is the one process of those being considered that is most highly dependent on oxygen. In addition to the usual oxygenation steps for phenylpropanoid elaboration, the subsequent oxyradical formation (Gross, 1980; Gross, Janse, and Elstner, 1977; Halliwell, 1978) involves the transformation of oxygen to superoxide. The disproportionation of superoxide gives hydrogen peroxide, which is utilized (via peroxidase) to yield the key oxyphenylpropanoid radical. The subsequent polymerization appears to be nonenzymatic. Presumably lignin formation could not proceed and could not have developed in the presence of low O_2 partial pressures. It is a reasonable to assume, from the earlier discussions on cutin and phenylpropanoids, that partial pressures of O_2 would need to be in excess of 0.004 atmospheres, or 0.02 PAL O_2, for lignin formation. One of the more interesting features of lignin formation is the involvement of superoxide dismutase. Although the nature of the superoxide dismutase (SOD) involved in the reaction is unknown, it is tempting to speculate that it is the cytosol Cu/Zn form rather than the mitochondrial Mn/SOD or Fe/SOD. The Cu/Zn SOD is distributed between higher land plants, charophycean green algae (*Nitella, Spirogyra*), fungi, and animals — a distribution that admittedly poses some interesting phylogenetic problems. It has not been identified in prokaryotes, other algae, or green algae of the chlorophycean or ulvaphycean lineage (see Chapman and Schopf, 1983, and citations therein). It seems reasonable to speculate, certainly from a basis of comparative biochemistry, that the appearance of Cu/Zn superoxide dismutase appeared as a response to rising O_2 levels and a need to protect against internal oxygen toxicity from direct exposure to atmospheric O_2 in place of dissolved O_2. In the case of land plants the protective mechanism — and its byproduct, hydrogen peroxide — has been incorporated into the synthesis of lignin. I would suggest that the oxygen requirements for lignin biosynthesis (in terms of partial pressure) must be at least as high, if not higher, than those for cutin formation and general phenylpropanoid metabolism and elaboration. If this is the case then one can reason that the early transitional or semiaquatic plants were perhaps incapable of lignin biosynthesis. The increasing oxygen partial pressures (which continued to increase after the migration) subsequently allowed for lignin biosynthesis; hence the vascularized land plants could have evolved only after the initial migration. This scenario is, of course, dependent upon the assumption of a late date for the appearance of O_2 levels in the range 0.01–0.1 PAL.

Evolutionary Position of the Bryophytes: Nature of the Migratory Land Plant

Early views often held that bryophytes, because their simplified morphology and partial adaptation to a terrestrial habitat, represented an early *ancestral* land-plant stock. The current prevailing opinion (Remy, 1982; Schuster, 1979; Smith, 1978), as illustrated in figures 3.1 and 3.2, holds that bryophytes and tracheophytes diverged from a common stock. This hypothetical ancestral group may thus represent the transitional flora. The nature of these organisms, especially with regard to their biochemistry (for example, the formation of lignins and cutins), pertinent to water supply and retention is critical to arguments developed in this chapter. It has been pointed out (Miller, 1970; Raven, 1977) that a *successful inhabitation* of the land depended upon an internal water-conducting system. The fossil and paleobiochemical record of the vascular land plants, bryophytes (offshoots of the transitional or first semiaquatic plants, which were cutinized but not vascularized), and the ancestral stock to both of these (the first semiaquatics?; Raven, 1977) is critical to time correlations.

Other Hypotheses

It is not my intention to suggest that an oxygen link was the *prime* control of the timing of the invasion of the land, only that it may well have been a component, provided that certain assumptions are made. Oxygen, or rather the lack of it, may have acted as a restraint on a potential land migration and development of vascularized plants. In recent years a number of other hypotheses have been discussed. The UV-phenylpropanoid connection has already been considered (Lowry, Lee, and Hebant, 1980) as an essential feature of a land migration. As this paper reemphasizes, UV protection, which is linked to O_2 levels through O_3 levels, may well have been a critical factor. Raven (1977) has recently considered the problem from the point of view of maintaining a homoiohydric condition (that is, the capacity to remain hydrated under circumstances of limited water supply). There is no question that the development of such a capacity was essential for the appearance of land plants. The evolution of lignin and cutin is an integral feature for the acquisition of this homoiohydric state. Raven has emphasized this and points out that some green algae show preadaptation along these lines. In a more provocative article, Lewis (1980) has suggested a key role for borate in the hydroxylase and oxidase steps in lignin biosynthesis. The origin of this role depended upon the selection for sucrose (a weak complexer with borate), rather than acyclic polyols, which bind and sequester borate, in the Chlorophyta. Without dismissing Lewis's hypotheses, one must remember we know very little about the function of borate in plant cells. If green algae (Charophyceae) were preadapted for phenylpropanoids and had selected for sucrose biosynthesis, one can ask why there has not been a greater elaboration of C_6-C_3 metabolism in this group. Borate is certainly abundant in aquatic systems.

Semiaquatics and presumably the ancestral migratory flora in moist environments are not subject to the potential problems of nutrient deficiency that plague strictly terrestrial plants. The role of mycorrhizal symbioses in the subsequent adaptation of semiaquatics and the development of a truly terrestrial flora is probably very important and reflects the need of a terrestrial plant to overcome the problems of nutrient shortage.

In the final analysis, however, the land migration and subsequent terrestrial development surely reflect an aggregate of a number of factors, some of which may have been environmental.

Proof and Conjecture

In this chapter oxygen, acting as a restraint mechanism in a number of ways, has been proposed as a contributing factor in the evolution of a land flora. The suggestions, however, depend upon a knowledge of the rate of increase of O_2 in the atmosphere and the time frame. At the biochemical level one is interested in two O_2 levels, approximately 0.002

PAL and 0.02 PAL; for the attainment of an external ultraviolet screen the level is 0.1 PAL. Calculations, on one hand, have indicated that the earliest *possible* date for 0.1 PAL O_2 was 1.7 Ga ago (Walker et al., 1983; cf. Canuto et al., 1982), whereas paleobiological data (Rhoads and Morse, 1971), on the other hand, have been used to estimate a much later date, 700–800 million years ago for 0.01 PAL and 570 million years ago for 0.1 PAL. If the former estimates are the more accurate, then hypotheses involving UV screens or the role of O_2 in biosyntheses such as outlined here are probably null and void. There is no fossil evidence for a land flora or land transition at that early date. If the latter dates are the more reasonable, then the hypotheses outlined in this paper should be considered; these theories are illustrated in figure

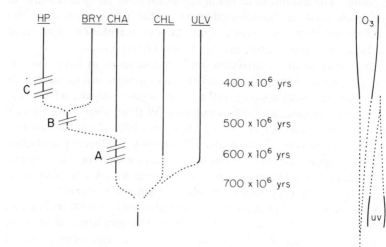

HP: HIGHER PLANTS: VASCULAR
BRY: BRYOPHYTA
CHA: CHAROPHYCEAE
CHL: CHLOROPHYCEAE
ULV: ULVAPHYCEAE

Fig. 3.4. Schematic diagram relating evolution to the biochemical and environmental changes, as suggested by the "late scenario" for O_2/O_3 appearance. The change in width of the O_3 and ultraviolet radiation bars is relative and is designed to represent a possible time frame of appearance and disappearance only.
  A. Appearance of Cu/Zn superoxide dismutase and phenylpropanoid elaborations.
  B. Appearance of cutin.
  C. Appearance of lignin and suberin.

3.4. In this regard I quote Kasting and Donahue (1980, 3255) in their introduction: "A question of particular interest is whether or not the development of the ozone layer was responsible for the emergence of life on land during the late Silurian period, about 420 m.y. (million years) ago." These authors conclude that "the uncertainty inherent in both calculations [of earliest and late limits], however, leaves open the possibility that the ultraviolet shield developed after the Cambrian was directly linked with the spread of life onto land" (3261). It is also important to remember that the fossil record pertaining to the land transition, which is sparse and in many instances difficult to interpret, is not a promise that a particular event occurred at a particular time. More important, though, is the question, can these hypotheses be tested? A number of experimental observations could provide data for their validity. These should include investigations on the effect of O_2 partial pressure between 0.001 PAL and 0.1 PAL on the activity of key enzymes in lignin biosynthesis (cinnamic hydroxylase, tyrosinase, peroxidase), superoxide formation, and aggregate lignin biosynthesis; in cutin biosynthesis (fatty acid dehydrogenase, hydroxylase, and epoxidase) and total cutin biosynthesis. Geochemically based and theoretical calculations of O_2 partial pressures and the rate of increase between one Ga and 300 million years ago are needed. To avoid the pitfall of a circular argument, these must be independent of any biological input. Gaps in our knowledge of comparative biochemistry need closing. Specifically, we need to look to the bryophytes and Charophyceae with respect to Cu/Zn superoxide dismutase, peroxidase, phenylalanine ammonia lyase, cinnamic hydroxylase, and general phenylpropanoid metabolism. The solution to such fundamental evolutionary questions as the origin of the land flora will come only from a multidisciplinary approach with input from a variety of sources.

Summary

The origin of a land flora is discussed from a number of viewpoints. A distinction is made between the migration to land (resulting in semi-aquatic plants) and the development of a terrestrial flora (that with lignin for vascularization and cutin for waterproofing). The role of oxygen plays a central part in the discussion, which includes a consideration of the development of an ozone atmosphere and UV shield and the possible need for early land plants to protect against UV irradiation. Following Lowry, Lee, and Hebant (1980), I propose that dependence on O_2 for the synthesis of phenylpropanoid (C_6-C_3) compounds played a part in the timing of the invasion of the land. I suggest that the

O_2-dependent biosynthesis of lignin (for mechanical support and water transport) and cutin (for waterproofing and shielding from toxic solutes) may be related to O_2 partial pressures. Time calculations for the origin of a land flora and the preceding migration are derived from biochemical-geochemical data. The timing of these events compares well with the time frame suggested by the fossil record. I propose that the changing environment, and particularly changes in O_2 partial pressures, played a key part in the migration of the Charophyceae to land and the subsequent development of a land flora.

Acknowledgments

I am indebted to Professors J. W. Schopf and D. B. Walker for their comments on the manuscript, and to the reviewers, Professors A. Knoll, K. Niklas, and B. Tiffney, for their in-depth reading and criticisms, especially with regard to paleobiological matters. To all these people I express my thanks for their unbiased and critical appraisals.

References

Banks, H. P. 1975a. Early vascular land plants: Proof and conjecture. *Bioscience* 25:730–37.

———. 1975b. The oldest vascular plants: A note of caution. *Rev. Palaeobot. Palynol.* 20:13–25.

Bower, F. O. 1908. *The origins of a land flora.* London: Macmillan.

Chapman, D. J., and J. W. Schopf. 1983. Biological and biochemical effects of the development of an aerobic atmosphere. In *Origin and evolution of the earth's earliest biosphere*, ed. J. W. Schopf, 302–20. Princeton: Princeton Univ. Press.

Edwards, D., M. G. Bassett, and E. C. W. Rogerson. 1979. The earliest vascular land plants: Continuing the search for proof. *Lethaia* 12:313–24.

Gorham, J. 1977a. Lunularic acid and related compounds in liverworts, algae and *Hydrangea*. *Phytochem.* 16:249–53.

———. 1977b. Metabolism of lunularic acid in liverworts. *Phytochem.* 16:915–18.

Grambast, L. J. 1974. Phylogeny of the Charophyta. *Taxon* 23:463–81.

Gray, J., and A. J. Boucot. 1977. Early vascular land plants: Proof and conjecture. *Lethaia* 10:145–74.

Gross, G. G. 1980. The biochemistry of lignification. *Adv. Bot. Res.* 8:25–63.

Gross, G. G., C. Janse, and E. F. Elstner. 1977. Involvement of malate, monophenols, and the superoxide radical in hydrogen peroxide formation by isolated cell walls from horseradish (*Armoracia lapathifolia* Gilib). *Planta* 136:271–76.

Halliwell, B. 1978. Lignin biosynthesis: The generations of hydrogen peroxide and superoxide by horseradish peroxidase and its stimulation by manganese (II) and phenols. *Planta* 140:81–88.

Hebant, C. 1979. Conducting tissues in bryophyte systematics. In *Bryophyte systematics*, ed. G. C. S. Clarke and J. G. Duckett, 365–83. London and New York: Academic Press.

Herak, M., V. Kochansky-Devide, and I. Gusic. 1977. The development of the Dasyclad algae through the ages. In *Fossil algae*, ed. E. Flugel, 143–53. Berlin: Springer-Verlag.

Jeffrey, C. 1962. The origin and differentiation of the archegoniate land-plants. *Bot. Notiser* 115:446–54.

Kasting, J. F., and T. M. Donahue. 1980. The evolution of atmospheric ozone. *J. Geophys. Res.* 85:3255–63.

Kolattukudy, P. E. 1980. Biopolyester membranes of plants: Cutin and suberin. *Science* 208:990–1000.

―――. 1981. Structure, biosynthesis and biodegradation of cutin and suberin. *Ann. Rev. Plant Physiol.* 32:539–67.

Lewis, D. H. 1980. Boron, lignification and the origin of vascular plants — A unified hypothesis. *New Phytol.* 80:209–29.

Loeffelhardt, W., B. Ludwig, and H. Kindl. 1973. Thylakoid-gebundene L-Phenylalanin-Ammoniak-Lyase. *Hoppe-Seyler's Z. Physiol. Chem.* 354: 1006–12.

Lowry, B., D. Lee, and C. Hebant. 1980. The origin of land plants: A new look at an old problem. *Taxon* 29:182–97.

Malloch, D. W., K. A. Pirozynski, and P. H. Raven. 1980. Ecological and evolutionary significance of mycorrhizal symbioses in vascular plants (A review). *Proc. Natl. Acad. Sci. USA* 77:2113–18.

Markham, K. R., and L. J. Porter. 1969. Flavonoids in the green algae (Chlorophyta). *Phytochem.* 8:1777–81.

―――. 1978. Chemical constituents of the bryophytes. *Prog. Phytochem.* 5:181–272.

Melkonian, M. 1980. Ultrastructural aspects of basal body associated fibrous structures in green algae: A critical review. *BioSystems* 12:85–104.

―――. 1982. Structural and evolutionary aspects of the flagellar apparatus in green algae and land plants. *Taxon* 31:255–65.

Miksche, G. E., and S. Yasuda. 1978. Lignin of giant mosses and some related species. *Phytochem.* 17:503–04.

Miller, H. A. 1970. The phylogeny and distribution of the Musci. In *Bryophyte systematics*, ed. G. C. S. Clarke and J. G. Duckett, 11–40. London and New York: Academic Press.

Moestrup, O. 1978. On the phylogenetic validity of the flagellar apparatus in green algae and other chlorophyll a and b containing plants. *BioSystems* 10:117–44.

Niklas, K. 1976a. The chemotaxonomy of *Parka decipiens* from the Lower Old Red Sandstone, Scotland (U.K.). *Rev. Palaeobot. Palynol.* 21:205–17.

―――. 1976b. Morphological and ontogenetic reconstruction of *Parka decipiens* Fleming and *Pachytheca* Hooker from the Lower Old Red Sandstone,

Scotland. *Trans. — R. Soc. Edin.* 69:483–99.

———. 1976c. The role of morphological biochemical reciprocity in early land plant evolution. *Ann. Bot.* 40:1239–54.

———. 1980. Paleobiochemical techniques and their applications to paleobotany. *Prog. Phytochem.* 7:143–81.

———. 1982. Chemical diversification and evolution of plants as inferred from paleobiochemical studies. In *Biochemical aspects of evolutionary biology*, ed. M. H. Nitecki, 29–91. Chicago: Univ. of Chicago Press.

Niklas, K., and L. M. Pratt. 1980. Evidence for lignin-like constituents in early Silurian (Llandoverian) plant fossils. *Science* 209:396–97.

Niklas, K., and V. Smocovitis. 1983. Evidence for a conducting strand in Early Silurian (Llandoverian) plants: Implications for the evolution of land plants. *Palaeobiology* 12:126–37.

Niklas, K., B. H. Tiffney, and A. H. Knoll. 1980. Apparent changes in the diversity of fossil plants: A preliminary assessment. *Evol. Biol.* 12:1–89.

Pickett-Heaps, J. D. 1975. *Green algae: Structure, reproduction and evolution in selected genera.* Sunderland, MA: Sinauer Association.

———. 1979. Electron microscopy and the phylogeny of green algae and land plants. *Am. Zool.* 19:545–54.

Pirozynski, K. A., and D. W. Malloch. 1975. The origin of land plants: A matter of mycotrophism. *BioSystems* 6:153–64.

Pratt, L. M., T. L. Phillips, and J. M. Dennison. 1978. Evidence of non-vascular land plants from the early Silurian (Llandoverian) of Virginia, U.S.A. *Rev. Palaeobot. Palynol.* 25:121–49.

Pryce, R. J. 1972. The occurrence of lunularic acid and abscisic acid in plants. *Phytochem.* 11:1759–61.

Raven, J. A. 1977. The evolution of vascular land plants in relation to supracellular transport processes. *Adv. Bot. Res.* 5:153–219.

Remy, W. 1982. Lower Devonian gametophytes: Relation to the phylogeny of land plants. *Science* 215:1625–27.

Rhoads, D. C., and J. W. Morse. 1971. Evolutionary and ecologic significance of oxygen-deficient marine basins. *Lethaia* 4:413–28.

Schopf, J. W. 1970. Precambrian microorganisms and evolutionary events prior to the origin of vascular plants. *Biol. Revs. Camb. Philo. Soc.* 45:319–52.

Schuster, R. M. 1979. The phylogeny of the Hepaticae. In *Bryophyte systematics*, ed. G. S. C. Clarke and J. G. Duckett, 41–82. London and New York: Academic Press.

Smith, A. J. E. 1978. Cytogenetics, biosystematics and evolution in the Bryophyta. *Adv. Bot. Res.* 6:195–276.

Stebbins, G. L., and G. J. C. Hill. 1980. Did multicellular plants invade the land? *Am. Nat.* 115:342–53.

Stewart, K. D., and K. R. Mattox. 1975. Comparative cytology, evolution and classification of the green algae with some consideration of the origin of other organisms with chlorophylls a and b. *Bot. Rev.* 41:104–35.

———. 1978. Structural evolution in the flagellated cells of green algae and land plants. *BioSystems* 10:145–52.

Strother, P. K., and A. Traverse. 1979. Plant microfossils from Llandoverian and Wenlockian rocks of Pennsylvania. *Palynology* 3:1–21.

Swain, T. W. 1980. Flavonoids. In *Pigments in plants*, ed. F.-C. Czygan, 224–36. Berlin: Springer.

Taylor, T. N. 1982. The origin of land plants: A paleobotanical perspective. *Taxon* 31:155–77.

Walker, D. B. 1983. The response of epidermal cells. In *Compatibility responses in plants*, ed. R. Moore, 123–37. Waco: Baylor Univ. Press.

Walker, J. C. G., C. Klein, M. Schidlowski, J. W. Schopf, D. J. Stevenson, and M. R. Walter. 1983. Origin and environmental evolution of the Archaean–early Proterozoic earth. In *Evolution of the earth's earliest biosphere*, ed. J. W. Schopf, 260–90. Princeton: Princeton Univ. Press.

Wray, J. L. 1977. *Calcareous algae*, Amsterdam: Elsevier–North Holland.

RICHARD BEERBOWER

Early Development of Continental Ecosystems

Introduction

The fossil record demonstrates a radical change in the character of life in brackish, freshwater, and terrestrial habitats — categorized inclusively as *continental habitats* — in an 80-million-year interval from the Ordovician to the end of the Devonian. This "mega-event" involved not only the origin and diversification of several classes and even phyla of organisms but also extensive changes in the structure and dynamics of continental ecosystems. The most significant of these changes, which first appear in brackish water and extend progressively into fresh water and, finally, terrestrial (nonaquatic) ecosystems, include (1) elaboration and differentiation of trophic structure, (2) increases in productivity, and (3) shifts in dominant ecologic strategies (from resource conservation and mortality replacement to resource aquisition and competition).

Although many workers have speculated on the origin and early evolution of particular groups of nonaquatic organisms, little attention has been given to the development of freshwater or brackish-water ecosystems, and not much more to the structure and dynamics of early terrestrial ones. Olson's work (1966) on the trophic aspects of Permian freshwater and terrestrial communities suggests a starting point for such an analysis. Rolfe (1980) has briefly considered the development of trophic structure in Siluro-Devonian terrestrial communities; Hughes and Smart (1957) and Smart and Hughes (1972) as well as Kevan, Chaloner, and Savile (1975) have investigated coevolution of early

47

terrestrial arthropods and vascular plants; Edwards (1980) and Niklas, Tiffney, and Knoll (1980) have dealt with various aspects of the development of plant communities and adaptive patterns in early vascular plants. Denison's analysis (1956) of habitats occupied by early fish and Størmer's studies (1976) of early continental arthropods have provided important insights into the occurrence and composition of early continental ecosystems. In addition, considerable information can be extracted from the literature dealing with individual fossil taxa and particular floras and faunas and from published interpretations of coeval continental sedimentary rocks.

This chapter opens with a summary of our knowledge of early continental ecosystems and their structure, dynamics, and development. This synopsis will serve to substantiate the explanations of ecosystem development and of evolutionary events and trends that I put forward in the concluding section.

Pre-Silurian Continental Ecosystems

Assemblages and Biocoenoses

The pre-Silurian fossil record of continental biocoenoses is extremely sparse in spite of the widespread occurrence of sedimentary rocks representing continental habitats. Some fossil plant materials reported from late Ordovician near-shore deposits may represent emergent hydrophytes if not fully terrestrial plants (Gray and Boucot, 1977, 1978; Gray, Massa, and Boucot, 1982). Since their remains are known only from transported materials, definite assignment to a more specific habitat — littoral marine, brackish, freshwater, or terrestrial — is not possible. Thorough study of some marginal marine and estuarine deposits of mid-Ordovician age failed to yield macrophyte traces (Spjeldnaes, 1979; Fischer, 1978) and studies of Ordovician paleosols, although so far quite limited, argue against the presence of vascular plants or even of large, erect nonvascular ones in terrestrial habitats though they do not preclude the presence of microphytes or sprawling nonvascular macrophytes (Retallack, n.d.). Further, Cotter (1978) argues that the rarity of sinuous stream deposits in pre-Silurian continental sediments demonstrates the absence of rooted macrophytes, and Basu (1981) reaches a similar conclusion from the relatively great abundance of unweathered detrital potassium feldspar in these rocks (see below). The presence of a continental microflora, however, is indicated by the occurrence of stromatolites in deposits interpreted as brackish and freshwater (Hoffman, 1976; Hofmann, Pearson, and

Wilson, 1980) and by the similarity of some late Proterozoic microbial fossils to modern, subaerial prokaryotes (Golubic and Campbell, 1979). One may reasonably infer the presence of blue green and green algae and, in terrestrial habitats, lichens.

Evidence of continental faunas is equally sparse. Possible brackish-water deposits of mid-Ordovician age in Colorado yield medium-sized agnathan fish and medium- to large-sized arthropods (scorpions, eumalacostracans, aglaspids, synxiphosurans, and limulines — but, very strikingly, no euripterids) (Fischer, 1978). Spjeldnaes (1979) suggests that some Ordovician assemblages of agnathan fish and lingulid brachiopods were exposed to variable salinity. In addition, by analogy with similar occurrences in younger rocks, one might expect that some of the low-diversity, bivalve-lingulid assemblages in middle and late Ordovician deposits also represent brackish habitats. Finally, some shallow-water, near-shore trail-and-burrows assemblages ("Skolithos" and even "Cruziana" ichnofacies) may represent brackish-water biocoenoses (Fischer, 1978). Even if all these Ordovician assemblages are truly brackish-water, the abundance and diversity of the macrofauna, however, was clearly much lower in them than in comparable Siluro-Devonian assemblages (see below). The Ordovician brackish-water assemblages are not as clearly differentiated in taxonomic composition from marine assemblages as those in younger deposits.

At present, knowledge of freshwater and terrestrial animals is limited to burrows observed by Retallack (1981) in some late Ordovician paleosols. He attributes these to small arthropods, but the interpretation is not definite. One might reasonably expect the presence of such organisms as part of a soil microbiota; such organisms are very rarely fossilized, and their absence as body fossils is to be expected. Although early fish and euripterids have been interpreted as members of early freshwater ecosystems (Romer, 1955), all pre-Silurian (and most Silurian) fish occur in what appear to be marine or brackish-water assemblages (Denison, 1956; Spjeldnaes, 1979; Boucot and Janis, 1983). Further, Ordovician euripterids typically occur in marine or even hypersaline assemblages (Caster and Kjellesvig-Waering, 1964) and do not appear in brackish-water and freshwater ones until middle and late Silurian.

Habitats

Prior to the extensive deployment of macrophytes, continental habitats must have been very different from modern equivalents with respect to resources, physical disturbance, and uncertainty (that is, spatial patchiness and temporal instability). At present, macrophytes, particularly the tracheophytes, strongly influence the weathering and hydrologic regimes in continental environments and thus the patterns of erosion and deposition as well as the storage of water and release of mineralized nutrients. They not only control microclimates but also affect macroclimates. Finally, they are the dominant food sources for secondary production in most aquatic as well as terrestrial habitats.

The most important environmental effect of macrophytes involves differences in weathering rates and processes. First, the rate of chemical weathering would have been considerably lower without a cover of rooted (or rhizoid-bearing) plants because of (1) the low partial pressure of CO_2 in soil water due to a relative paucity of organic detritus, (2) the low level of fungal leaching due to the absence of root mycorrhiza and the paucity of detritus, and (3) the paucity of moisture resulting from rapid throughflow of soil water (see below). Moreover, the absence of deep root zones would have limited biologically mediated weathering to a near-surface film. Cawley, Burruss, and Holland (1969) found the rate of chemical weathering in central Iceland to be two or three times greater in areas covered by vascular plants than in those limited to lichens and bare soil. Basu (1981) presents evidence for low rates of weathering of potassium feldspar in the absence of plants, presumably due to the lack of direct uptake of potassium by plant roots. The same chemical situation would presumably hold for other critical nutrient ions, for example, phosphate, that are taken up selectively. It seems likely, therefore, that the overall release of nutrient ions from pre-Silurian, terrestrial soils was one-half or less than that at present (and much less if lichens were absent). This shortage would have limited productivity in continental habitats and affected nutrient availability in freshwater and marine systems.

Second, the physical character of soils probably differed on the average from modern ones. In particular, a lower rate of chemical weathering probably resulted in a lower production of clay minerals and a relative dominance of silt over clay in the fine fraction. Such a comparative shortage of fines would have been exaggerated by relatively high rates of splash and sheet erosion and of eolian deflation on unvegetated surfaces (except as opposed by algal and lichen crusts). Wind erosion would also have been intensified because of relatively high velocities close to the ground in the absence of plant windbreaks. Overall, the result of the absence of macrophytes would have been a relative predominance of sandy and rocky soils of high permeability but low porosity. Such soils would have had high infiltration capacity but low water retention, transforming surface runoff to rapid throughflow in the soil zone. Fine-grained, water- and nutrient-retentive soils would have been limited largely to lowland, alluvial or lacustrine areas.

The existence of such a physical regime would have had serious consequences for terrestrial plants. Essential ions for plant nutrition, held in modern soils on clay and organic aggregates, would have been removed very rapidly because of low retentivity and high water flux; only a small amount could have been retained in the algal or lichen crusts. Further, the upper soil zone would be quite dry except during and immediately after rainfall. Unvegetated soils would have dried rapidly, and surface moisture would have been quite unpredictable. Temperature variations would also have been very large.

The hydrology of terrestrial drainage basins would have differed because of a higher proportion of runoff and throughflow to interception and transpiration. Schumm (1968) suggests that in such circumstances the mean annual runoff would be 100 percent greater for a given annual precipitation than that observed in modern continental habitats. High, "flash" discharge implies a high rate of erosion aggravated by the absence of roots to check water flow and bind small particles, although algal crusts would reduce erosion somewhat. Such rapid erosion would have tended to maintain thin, rocky soils with very low storage capacity for either water or dissolved nutrients.

The physical and chemical characteristics of the streams would have reflected the differences in weathering, soils, discharge, and erosion. For a given climatic regime, stream water would have displayed relatively low ion concentrations (because of lower rates of chemical weathering and higher rates of discharge). Fine sediment trapped in the pools would have carried fewer absorbed ions because of the lower organic content (and, possibly, a lower clay fraction). Less fine sediment would be trapped because of the absence of aquatic macrophyte "baffles." Nutrient ions, rather than being accumulated by macrophytes and plant detritus and held for slow release and recycling, would have been lost rapidly. Finally, such nutrients as might be held in alluvial sediments would remain entombed rather than being pumped back into the ecosystem through the root systems of aquatic macrophytes (see Moss, 1980, 141, on the role of rooted macrophytes in modern freshwater ecosystems). Oxygen partial pressures in the water would vary primarily with temperature, since organic decomposition would consume relatively little oxygen; conversely, decomposition would tend to be rapid and thorough. Carbon dioxide partial pressures would tend to be far below saturation for the same reasons. The "flashy" hydrologic regimes would produce severe floods of quite brief duration, interspersed with periods of extremely low flow confined to channel thalwegs; stream waters would be very turbid and have a heavy traction load. These characteristics, together with highly variable discharge and lack of rooted bank vegetation would favor development of braided streams (Schumm, 1968). Channels and bars both would be subject to rapid change and be of impermanent character. (Cotter, as noted above, reports the predominance of braided, and rarity of sinuous, stream deposits in pre-Silurian sedimentary rocks.)

Lakes would also have diverged in character, although to a lesser extent than streams diverged, because of the differences of input associated with such extremes in weathering, erosion, and runoff as well as because of the absence of macrophytes. The lower input of dissolved nutrients in stream flow and the loosely bound nature of nutrients in sediments would have limited microphyte production. Additionally, the absence of macrophytes would have reduced nutrient accumulation in the ecosystem and, even more significant, reduced recycling from lake-bottom sediments. Thus oligotrophic lakes would have been proportionally much more abundant than at present, although some lakes presumably could have accumulated dissolved nutrients, given favorable local input/output ratios (at the cost, however, of being extremely alkaline and thus unfavorable for most organisms). Flood-basin lakes would have been relatively turbid, nutrient-poor, and subject to extreme changes in level and to extended periods of desiccation. Early lacustrine stromatolites such as the Proterozoic ones described by Hoffman (1976, 609–11) seem to represent such circumstances. Shallow, ephemeral ponds and backwater lagoons would have existed in the place of swamps and marshes; these might be enriched through accumulation of algal detritus, but would have been dried mudflats most of the year. In the absence of a leaf canopy, their temperatures would have fluctuated violently.

Estuaries and lagoons would have shown features similar to those of the lakes but with some further moderation due to the influence of ocean water. Input of dissolved nutrients and their accumulation and recycling in organic detritus would have averaged considerably below the modern rate. Even though seawater influx would tend to ameliorate such shortages (see Hopkinson and Day, 1979, and Fisher, Carlson, and Barber, 1982, on marine-continental balances in modern coastal ecosystems), there could have been no local recycling from buried sediments in the absence of rooted hydrophytes. One might also expect more marked variation in salinity as a result of the extreme variation in river flow — that is, brief episodes of extreme dilution alternating with periods of normal salinity (see Livingston and Duncan, 1979, on the effect of clear-cutting on modern estuaries and lagoons). Sediment transport, erosion, and deposition would also have differed because of the absence of macrophyte baffling and sediment binding in the littoral zone and because of the marked variation in sediment influx from the continent (marked, short-term alteration between high constructional and high destructional regimes).

Global environmental variables (atmospheric composition, ionizing radiation, paleogeography, and climatic zonation) do not appear to have exerted an unusual influence on early Paleozoic continental habitats or ecosystems. Carbon dioxide and oxygen partial pressures in the atmosphere are apparently regulated by biotic and chemical processes within the ocean (Garrels, Lerman, and Mackenzie, 1976; Walker, 1977). The near-constant distribution of marine communities since the late Cambrian argues for a partial pressure of oxygen much like that of the present, certainly within one order of magnitude (Rhoads and Morse, 1970; although Chapman takes an alternative view in this volume). If the oxygen partial pressure were more than 1 percent of the present level, then the flux of ultraviolet (UV) radiation could not have been significantly greater than at the present (Ratner and Walker, 1972); the presence of very shallow water marine communities in pre-Silurian sediments argues against high-intensity UV radiation at the ocean's surface.

Influence by paleogeographic factors is equally obscure. The distribution of particular continental units changed radically through the early and middle Paleozoic (Ziegler, Scotese, et al., 1979), but these changes probably had little effect on the nature and/or abundance of various continental habitats and thus did not direct phyletic evolution or ecosystem development. Even though one might postulate that geographic changes affected rates of species origination and extinction and produced unique ecologic interactions among previously isolated species, such effects would not seem *a priori* to demand any particular direction. (If, however, the development of continental ecosystems depended on extremely rare evolutionary events, the rate of speciation and the frequency of unique ecologic events during any particular interval would affect the likelihood of such events.)

Climatic indicators suggest polar as well as subtropical and tropical climatic regimes (Drewry, Ramsey, and Smith, 1974). The location of the zonal boundaries may have differed somewhat from the present due to differences in continental albedo and evaporation (plus transpiration) (Ziegler, Bambach, et al., 1981). On a global basis these differences would presumably reduce the importation of moisture into the subtropical zone from the adjacent equatorial and high-latitude belts and widen the subtropical arid/semiarid climatic belts (Charney, Stone, and

Quirk, 1975). This situation would increase the continental area subject to extreme uncertainty in rainfall and would have yielded a climatic pattern consistent with the occurrence of red beds at atypically low and high latitudes during this interval. In the absence of terrestrial macrophytes, rainfall patterns may also have differed from present patterns. Plant transpiration is a major factor in recycling water.

Extensive development of continental ice sheets in polar Gondwanaland during the late Ordovician must have been correlated with relatively major climatic fluctuations and episodic changes in sea level. The climatic variation was presumably similar to that in the Pleistocene (Ziegler, Bambach, et al., 1981, 234) and thus would have involved some latitudinal shifts in the boundaries of climatic zones and some expansion of the area of wet-dry and dry habitats. Whether sea-level changes (McKerrow, 1979) had any significant effect other than geographic displacement of coastal and lowland alluvial habitats is not clear, but theoretically eustatic regression might expose more of the gently sloping continental shelves and thus expand the area of such habitats. Pratt, Phillips, and Dennison (1978, 144–45) attribute late Ordovician– early Silurian alluviodeltaic deposits in the Appalachians to such an event. An increase in the extent of coastal lowland areas, if it occurred, might have increased the number of such relatively rich, favorable, and stable continental margin habitats. Such a development would have strong potential effects on both ecosystem development and organismal evolution (Gray and Boucot, 1978). On the other hand, paleogeographic reconstructions for the Cambrian and Ordovician show extensive lowland areas over most of the continental interiors (for example, Scotese et al., 1979), so there would have been no scarcity of such habitats in the early Paleozoic.

Overall, as noted by numerous authors (Swain and Cooper-Driver, 1981; Gray and Boucot, 1977), pre-Silurian continental habitats were marked by limited resources of inorganic nutrients and water, by severe physical stresses that would have interfered with resource utilization, and by extreme physical patchiness and temporal instability. Deductively, one would expect these constraints to be most severe in terrestrial habitats, particularly in continental interiors and uplands, and least severe in large, permanent delta lakes, brackish lagoons, and estuaries.

Ecosystems

With such limited evidence from fossils and rocks, the reconstruction of pre-Silurian ecosystems primarily depends on deduction from modern analogues and from general ecologic principles. The scarcity of fossils suggests that most or all of the organisms were small and that any large ones lacked lignified or mineralized structure and/or were extremely rare. One can also exclude groups whose evolutionary origins are known to be post-Ordovician. Finally, the habitat characteristics (inferred from the absence, or near absence, of macrophytes and from the sedimentologic evidence) set limits on ecosystem structure and dynamics as well as on individual adaptions.

In terrestrial habitats, the limited availability of water as well as slow release and low retention of inorganic nutrients would have kept primary productivity to a very low level except in the most favored sites. High physical stress and uncertainty would have further limited production to brief time intervals in small, scattered sites. The primary producers were predominantly microphytes, with macrophytes contributing very little, if anything. Secondary production in the micro- and meiofauna must also have been limited in amount and in temporal and spatial distribution. These faunas presumably included a mixture of microherbivores and decomposers (representing a second trophic level) plus meiodetritivores and meiopredators (the third trophic level) not unlike those in modern dry-soil biotas (Rolfe, 1980; the burrows reported by Retallack, 1984, may represent such organisms). The lack of fossils representing a macrofauna suggests the absence, or at least the extreme rarity, of animals at the fourth trophic level (except parasites). The typical structure of these early terrestrial ecosystems would thus have comprised three trophic levels with a decomposer loop.

Freshwater ecosystems would also have differed from modern analogues, though perhaps less sharply than did the terrestrial ones. In general, stream ecosystems would have been subject to severe disturbance with unpredictable variation in discharge and with episodic, rapid erosion and deposition. At present, such ecosystems are highly dependent on allochthonous organic matter and on autochthonous production by rooted, semiaquatic macrophytes (Moss, 1980; Welcomme, 1979). Neither planktonic nor benthic microphytes contribute much to the food base except where eutrophication is extreme or where they accumulate in the semiponded, lower reaches of large streams (Roberts, 1972). With very low release of dissolved nutrients from the land by weathering, and without nutrient retention and recycling by rooted plants, microphyte productivity would have been even lower in pre-Silurian ecosystems than in modern analogues. One may reasonably assume (as discussed above) the presence of decomposers and a fauna of small animals including micro- and meioherbivores, meiodetritivores, and meiopredators. Again the lack of macrofossils suggests that consumers representing the fourth trophic level were absent, though one cannot rule out some opportunistic macropredators, for example, large arthropods, invading from brackish habitats. Pre-Silurian aquatic ecosystems are further differentiated from later ones by the absence of the macrodetritivores, for example, some arthropods as well as agnathan fish, which are known from Ordovician brackish-water deposits as well as from middle Silurian and younger alluvial ones.

Autochthonous primary production in lacustrine habitats would also have been restricted by low input of both dissolved and organically bound inorganic nutrients and by the lack of rooted plants to recycle such nutrients after burial in bottom sediments. While input of allochthonous organic detritus would have been relatively small, lakes should have been more productive on the average than rivers and supported at least moderate algal productivity and standing crop. Physical conditions, though fluctuating more than those in modern equivalents, would still have been more stable than in coeval stream or terrestrial habitats. Here again the presence of micro- and meiofaunas at three trophic levels may be presumed, but the lack of macrofossils suggests the complete absence of a fourth trophic level and of macrodetritivores.

The latter situation may, however, be more apparent than real; Ordovician lacustrine habitats are inadequately sampled because of (1) relatively limited investigation of pre-Silurian continental deposits; (2) dominance of piedmont plain, braided-stream deposits in some of the more thoroughly studied units — for example, the Juniata Sandstone of the central Appalachians (Cotter, 1978) — a regime not likely to include either permanent flood-basin or deep-basin lakes; (3) rarity of large animals in unproductive habitats; and (4) the difficulty in finding pre-Silurian fish fossils, which consist almost entirely of small, obscure, isolated tubercles or spines (Spjeldnaes, 1979). Thus, although Denison (1956) has argued persuasively for a marine origin for the vertebrates and for the marine origin of pre-Ludlovian fish, his evidence and interpretations are not incompatible with a limited penetration of fish into delta-plain lacustrine habitats during the Ordovician. Certainly the classic occurrences of fish in the Harding Sandstone appear to include sites representing at least brackish-water habitats.

Brackish lagoons and the estuarine portions of large, sluggish rivers would have resembled large lakes except for (1) the potential addition of dissolved nutrients through seawater mixing and (2) the disturbance associated with tides and oceanic storms (Fisher, Carlson, and Barber, 1982). In many modern analogues, rooted macrophytes (for example, mangroves or salt grass) are the dominant producers and also enhance microphyte productivity by recycling mineralized nutrients locked in bottom sediments. Without rooted macrophytes, the principal production would presumably be in the marine-dominated portions of these habitats. Some Ordovician faunal assemblages (Spjeldnaes, 1979) appear to record habitats that had near-normal salinity much of the time and therefore were subject to frequent colonization by marine organisms.

The only strictly brackish-water ecosystem known from the Ordovician seems to be that reported by Fischer (1978) from a few localities in the Harding Sandstone in Caloradic, which have yielded a rather small number of fossils. Trophic structure of the ecosystems represented by these occurrences is obscure, but the absence of any macrophyte traces suggests the predominance of microphytes as producers; the fish and some of the arthropods were probably deposit feeders (detritivores), and the lingulids were suspension feeders. Some of the arthropods may have been macropredators, but since they preyed primarily on the large deposit feeders (effectively at the second tropic level), they may be regarded as representing the third rather than fourth trophic level. In any case, the deposit feeders appear to show much greater variety than suspension feeders and predators. Though a single occurrence can hardly support very precise conclusions about average conditions, the general Ordovician situation contrasts markedly with that in the Devonian, where "upper-estuarine" assemblages are common, definitely include a fourth tropic level, and produce relatively large numbers of individuals (see discussion below). On the other hand, the occurrence of several groups of arthropods as well as fish in brackish habitats demonstrates that these forms had already adapted to low-salinity environments and that the paucity of such occurrences in the Ordovician (and early Silurian) cannot be attributed to tardiness in evolution of osmoregulatory capacities. This suggests in turn that the absence of arthropods and fish from Ordovician and early Silurian freshwater biocoenoses results from some environmental constraint other than low salinity.

Siluro-Devonian Assemblages, Habitats, and Ecosystems

Overview

The number and the distinctiveness of continental assemblages increased through the Silurian and Devonian. This sequence resolves, at least in Europe and North America, into four phases:

1. Llandoverian and Wenlockian Ages (early and middle Silurian, approximately 445–430 million years before the present);

2. Ludlovian and Pridolian (middle and late Silurian, 430–415 m.y.b.p.);
3. Gedinnian, Siegenian, and Emsian (early Devonian, 415–395 m.y.b.p.);
4. Eifelian, Givetian, Frasnian, and Famennian (middle and late Devonian, 395–370 m.y.b.p.).

This sequence demonstrates fundamental changes in the characteristics and distribution of continental biocoenoses in spite of some biases induced by the relative paucity of Silurian continental deposits in the areas of intensive investigation (in North America and western Europe). Because of the scarcity of Silurian deposits, the changes between early and late Silurian may be more apparent than real; the "first appearance" of a group in late Silurian or very early Devonian may not preclude an earlier origin, such as the middle or early Silurian. The effects of such potential biases are considered in my interpretation of these phases.

Alluviodeltaic deposits (representing the various aquatic and non-aquatic continental habitats) were particularly widespread from late Silurian through the Devonian in the Laurentian, or "Old Red," continental unit of northwestern Europe, Greenland, and eastern and arctic North America. This obviously has provided the opportunity for a more complete knowledge of continental biocoenoses (as noted above) for this time interval and area. However, discrepancies between the "Old Red" record and that in Australia and in North Africa suggest that the former is not fully representative. These discrepancies are discussed below.

Llandoverian-Wenlockian

Various plant and animal fossils have been ascribed to early and middle Silurian continental ecosystems, though neither age nor habitat have been determined precisely. The oldest traces of likely continental macrophytes appear to be those (1) in alluvial deposits of Virginia (which are probably Llandoverian though they might be as late as Ludlovian; Pratt, Phillips, and Dennison, 1978), (2) in marine deposits of similar or greater antiquity from Maine (Schopf et al., 1966), and (3) in the alluvial Acacus Formation of Libya (which may be late Llandoverian though it could be Wenlockian or even as late as Gedinnian, per Massa, 1980, and Douglas and Lejal-Nicol, 1981). In addition, remains of macrophytes and of possible continental macrozoans occur in Scottish rocks that might be as old as Llandoverian (Lamont, 1948) but are more likely of Wenlockian or even Ludlovian age (Westoll, 1950; Denison, 1956); they apparently represent deposition in transitional marine-to-freshwater habitats (Denison, 1956; Waterston, 1979).

The plant materials include macrophytic algae, probably chlorophytes, from Scotland (Lang, 1937; Niklas, 1976b, 1980), nematophytes from Virginia and Scotland (Pratt, Phillips, and Dennison, 1978; Lang, 1937), and tracheophytes from Libya (Boureau, Lejal-Nicol, and Massa, 1978) and, possibly, from Maine (Schopf et al., 1966). The chlorophytes are almost certainly aquatic, probably freshwater, and possibly brackish-water but interpretation of the nematophytes is somewhat more problematic. Although Lang argues for a fully terrestrial habit, the evidence does not require such a conclusion (Edwards, 1980). The occurrence of the nematophytes in the rocks requires some to have been nonmarine (though not necessarily nonaquatic), but it does not exclude the possibility that some were brackish-water or even marine plants (Smith, 1979). Their morphology (Lang, 1937; Pratt, Phillips, and Dennison, 1978) and their geobiochemistry (Niklas, 1976a; Niklas and Pratt, 1980) imply adaptation to subaerial conditions but possibly as emergent hydrophytes and/or as littoral and riparian plants (Gray and Boucot, 1977, 1978). The argument for extensive occupation of terrestrial habitats is further weakened by lack of attachment to any sort of basal absorptive organs and by the absence of scleritized tissues necessary to support an erect habit.

The presumed macrophytes from Maine consist of small, simple, erect axes and occur in strata that otherwise bear near-shore assemblages of marine fossils (Schopf et al., 1966). Although Strother and Lenk (1983) argue that they represent worm burrows, they possess a coalified outer layer suggesting thick-walled cells, and their chemical composition resembles that of some late Silurian and early Devonian tracheophytes (Niklas, 1980). The form of the axes resembles that of early tracheophytes, so they might be regarded as representing "proto-tracheophytes" (Edwards, Bassett, and Rogerson, 1979; Raven, 1977) though lack of preserved microstructure prevents a test of this suggestion. Their habitat was most likely littoral marine though neither sublittoral nor supralittoral interpretations can be excluded.

The plant fossils from Libya clearly represent tracheophytes, though small and relatively simple ones. They are clearly adapted for terrestrial as well as semiterrestrial habits, but their geologic age is in considerable doubt, as noted above. They occur in otherwise unfossiliferous continental deposits that overlie Middle Llandoverian marine deposits and underlie Siegenian marine units; they might be the lateral time equivalents of Late Llandoverian and Wenlockian marine deposits present in a separate basin to the north. The plants consist, however, of genera characteristic of Siegenian and post-Siegenian strata in Europe and North America, although forms of similar evolutionary grade have been assigned a Ludlovian date in Australia (Douglas and Lejal-Nicol, 1981). In terms of my analysis of ecosystem development, the most conservative viewpoint would seem to accept the presence of a few small, primitive tracheophytes in early Silurian continental biocoenoses.

In addition to plants, the mid-Silurian continental rocks of Scotland have yielded what is probably a brackish-water fauna with an admixture of possible freshwater and terrestrial elements (Denison, 1956; Waterston, 1979). The assemblages include ostracods, eurypterids, limulids, scorpions, myriapods, and fish in addition to lingulid brachiopods. The ostracods represent genera that appear elsewhere in transitional marine-nonmarine situations; the limits on their distribution are uncertain but likely extend to brackish-water habitats (Hoskins, 1961). Some of the eurypterids may have been amphibious (for example, *Drepanopterus*, per Størmer, 1976, and Rolfe, 1980, though Watenston, 1979, suggests they were predominantly aquatic), and several of the genera represent the "brackish phase" eurypterids of Kjellesvig-Waering (1961). The scorpions show no trace of book lungs and may have been gill-bearing (Laurie, 1899); like other known pre-Carboniferous scorpions they were probably aquatic or semiaquatic (Størmer, 1976). The myriapods are conventionally considered to be terrestrial, though Størmer (1976) suggests that an early Devonian one from Germany might have been amphibious. The fish consist entirely of primitive, jawless types — agnathans, including anaspids, thelodonts, and osteostracians. In the late Silurian these groups typically appear in brackish-water assemblages, although some occurrences may represent freshwater habitats or at least include individuals transported from such situations (Denison, 1956).

The evidence from early and mid-Silurian sedimentary rocks thus demonstrates the presence and some of the general features of continental ecosystems. Macrophytes were probably present as producers in both aquatic and terrestrial ecosystems and probably included forms with rhizome-rhizoid systems for water and nutrient collection. At the very least, the presence of macrophytes would have tended to stabilize the amount of living biomass and of detritus available to animals and decomposers, as in modern ecosystems (Cummins, 1975). They would have increased primary production by drawing on nutrients and water from the substrate, as well as by utilizing the greater intensity of light available to emergent hydrophytes and subaerial plants. To the extent that they utilized terrestrial habitats, they would have supplied allochthonous detritus to aquatic habitats and increased the rate of nutrient release through accelerated weathering. Rhizome mats associated with littoral, riparian, and terrestrial macrophytes could have damped episodic erosion and deposition to some extent and might have reduced variations in runoff and throughflow.

These deductions are consistent with changes observed in the consumer components of continental ecosystems. Brackish-water assemblages are far more common in the early and middle Silurian than in the Ordovician and include a greater abundance and diversity of animals, among them meiodetritivores (ostracods and, possibly, myriapods), meiopredators (scorpions), and macrodetritivores (limulids and some variety of agnathan fish). In addition, the eurypterid invasion of brackish water added macropredators (though some eurypterids were detritivores, per Waterston, 1979, 316–17). Three trophic levels and a major decomposer-detritivore loop are clearly distinguishable; the scorpions, which are large enough to have fed on other (unfossilizable) meiopredators, may represent the fourth level, as may some eurypterids. Many of the latter, however, were probably largely dependent on macrodetritivores and were thus effectively of the third, rather than the fourth, trophic level. Parallel though less extensive changes had occurred in freshwater ecosystems if, as Denison (1956) suggests, some of the agnathan fish actually lived in freshwater habitats. In addition, some of the myriapods and scorpions may also represent freshwater ecosystems — the former along with some fish filling the second trophic level (as detritivores) and the latter the third level (as predators on the meiobiota). Terrestrial biocoenoses, though very poorly known from lower Silurian fossil assemblages, probably included some small vascular plants (tracheophytes) as well as various microphytes. Some (or all) of the millipedes represent the second trophic level and, if some of the scorpions were amphibious, they would have operated at the third level as meiopredators. The absence or rarity of small arthropods in early freshwater and terrestrial assemblages has little positive significance since the probabilities of their fossilization and discovery are diminishingly small. Only three significant occurrences have been reported from the Devonian under much more favorable circumstances for discovery. Algal crusts could have supported some abundance and diversity of such forms and their occurrence in two early Devonian assemblages (see below) suggests their presence and differentiation by late Silurian if not considerably earlier (Rolfe, 1980).

On the other hand, any increases in productivity and reductions in stress and uncertainty in continental ecosystems could not have been very great. Carbonaceous accumulations (coaly lenses and carbonaceous shales) remained very rare in continental sediments until the late Devonian. The continued relative dominance of braided-stream deposits into Devonian time (Cotter, 1978) speaks for persistence of flashy stream discharge, for abrupt, erratic erosion and deposition, and thus for general instability in terrestrial, freshwater, and even brackish-water habitats. The paucity of continental fossil assemblages also points toward low productivity (if not high stress and uncertainty as well).

Ludlovian-Pridolian

The best-known late Silurian continental assemblages occur in the Welsh borderland of the British Isles and are supplemented by occurrences in Scotland, Norway, Spitsbergen, and eastern North America. The Libyan localities discussed above and those in Australia which yield the "Baragwanathia flora" may also date to the late Silurian (Garrett, 1979; Boucot and Gray, 1982). Although dominated by deposits representing marine to brackish-water transitional habitats, the known assemblages do include some upper-estuarine and freshwater situations. Unfortunately, the latter are relatively late (Pridolian), so evolutionary sequences cannot be detailed and the Ludlovian-Pridolian interval must be treated as a whole.

Although interpretation of the producer components is confused by the disputed dating of the early fossil plants from Libya and Australia, it is accepted that tracheophytes appeared at least as early as Ludlovian, as attested by bundles of tracheids and by small tracheophyte-like axes from the British Isles and eastern North America (Edwards and Davies, 1976; Edwards, Bassett, and Rogerson, 1979). These plants were all quite small with axial diameters of about 3 mm, had a simple morphology with leafless, dichotomous branches (Chaloner and Sheerin, 1979), and are assigned to the rhyniophytes. Their basal portions are unknown, but rhizomes and rhizoids appear on similar axes from Gedinnian deposits and thus are presumed to have been present in late Silurian forms (Edwards, 1980). The Libyan and Australian occurrences demonstrate the existence of other tracheophytes, specifically the trimerophytes and the lycopsids. The former are distinguished by multiple, unequally branching, leafless axes. The lycopsids are relatively complex forms marked by elaboration of branching and by development of vascularized leaves arising from the surface of the branches. These features would have increased the effective area for photosynthesis and simultaneously required increased nutrient and water uptake from the soil. Although all these fossils derive from near-shore marine and adjacent continental margin deposits, they probably reflect the composition of continental ecosystems as a whole since they provide a sufficient source for the limited variety of trilete spores dispersed through these same deposits; there is no evidence of "exotic" spores transported in from different types of upland plants (Chaloner, 1972; K. C. Allen, 1981).

Nematophytes and algae (see preceding section) occur in considerable numbers in many of the late Silurian continental assemblages, along with remains of early tracheophytes. Lang (1937) interpreted these assemblages as terrestrial but, since they represent plant fragments transported to their burial sites, they may be an admixture of aquatic and nonaquatic species derived from freshwater or even brackish habitats as well as terrestrial (Edwards, 1980). The first fossil charophytes, a group whose modern representatives are all submerged, freshwater macrophytes, also appear in these deposits (Croft, 1952).

Brackish-water fossil assemblages include the same major animal groups known from the middle Silurian (see discussion of mid-Silurian inliers of Scotland above) plus some significant additions. Holdovers include lingulid brachiopods, ostracods, eurypterids, limulids, and possibly scorpions among the arthropods, and anaspids, thelodonts, and osteostracians among the agnathan (jawless) fishes (Denison, 1956, 372–74; Allen and Tarlo, 1963; Kjellesvig-Waering, 1961). In addition, some bivalves and snails occur in what seem to have been brackish-water situations (Allen and Tarlo, 1963), and heterostracian agnathans apparently began to invade brackish (and freshwater) habitats by Pridolian time (Denison, 1956, 419–20; Allen and Tarlo, 1963). Finally, and probably most significant from both ecologic and evolutionary viewpoints, primitive jawed fish (the acanthodians) make their first appearance in brackish assemblages at least by Pridolian time (Allen and Tarlo, 1963).

By late Silurian, freshwater fossil assemblages are somewhat more easily distinguished from brackish-water ones. Indeed, Allen and Tarlo (1963) recognize several distinct assemblages in the lagoon-estuary-river-lake habitat spectrum of the Old Red Sandstone in the Welsh borderland. In spite of this increased differentiation, most of the major groups are the same as those in brackish-water assemblages. These probably include some ostracods, eurypterids, and limulids as well as scorpions (see Allen and Tarlo, 1963, and Denison, 1956, 372–74, *contra* Kjellesvig-Waering, 1961) and may also include some myriapods (Størmer, 1976). The osteostracian and anaspid agnathans also continue in freshwater biocoenoses from mid-Silurian time, albeit somewhat more diverse and more widely distributed than in late Devonian deposits. Newcomers include heterostracian agnathans (previously predominantly or entirely marine) and primitive jawed fish, the acanthodians. In addition, freshwater biocoenoses very possibly contained a variety of small arthropods that have not been observed so far (except for ostracods) but do appear, very rarely, in early Devonian assemblages.

Fully terrestrial faunas are very poorly known, though a few of the fossils preserved in aquatic assemblages were probably of terrestrial or semiterrestrial organisms. These consist of arthropods (first known from mid-Silurian occurrences), some fully terrestrial myriapods, and amphibious scorpions, eurypterids, and limulids. In addition, terrestrial biocoenoses probably included a meiofauna of small arthropods and various "worms" inhabiting the soil, algal crusts, and any litter accumulations generated by the now developing terrestrial macrophyte communities.

Overall, the most significant transformations in taxonomic composition between mid- and late Silurian continental ecosystems are (1) the expansion (or first appearance) of tracheophytes, (2) a modest diversification of brackish and freshwater fish, and (3) the appearance of large predators (jawed fish and possibly eurypterids) in freshwater habitats. The direct contribution of tracheophytes to primary production was probably limited by their small size and by the lack of well-differentiated foliage and true root systems. Further, they were probably quite limited in geographic and habitat distribution, as indicated by their rarity as fossils and by restriction of fossil occurrence to lower-alluviodeltaic deposits. In addition, within-habitat diversity was quite low (Edwards, 1980; Niklas, Tiffney, and Knoll, 1980). To the extent that the tracheophytes and, possibly, the nematophytes utilized terrestrial sites, they could have contributed to nutrient release in weathering, to retention of water in soils (Edwards, 1980, 75–76), to reduction of erosion, and to the supply of organic detritus for adjacent aquatic habitats. These groups plus the charophytes must also have been the major producers in aquatic habitats. By cycling mineralizing nutrients from the substrata, they may have wrought a modest increase in microphyte productivity as well. In all, primary production in continental ecosystems was almost certainly higher in late Silurian than in early Silurian.

The increase in abundance of animal fossils suggests an increase in secondary production, and the increase in variety of consumers demonstrates increased trophic diversity within brackish and probably freshwater ecosystems. The apparent increase in within-habitat diversities involved an increase in the number of distinct feeding types ("guilds"), as well as in the number of species within such guilds. Although the anaspid fish continued primarily as feeders on the nektonic meiofauna (Denison, 1956, 422–23, and 1961), with the appearance of cephalaspids and ateleaspids the osteostracians showed increased specialization for bottom-living, mud-grubbing habits (Westoll, 1979; Denison, 1956, 422). An increase in the variety of limulids and eurypterids also added to the diversity of large detritivores. The appearance of heterostracians and acanthodians in continental biocoenoses increased the diversity of forms feeding on the nektonic meiofauna (Denison, 1956, 419–20, 425–26, and 1961; Halstead, 1973). In addition, some of the acanthodians may have functioned as macropredators (Denison, 1956, 425, and 1979); these, together with some predatory eurypterids, may represent further differentiation of the fourth trophic level, though clearly this had not reached the level it was to attain by the end of the early Devonian.

The record of terrestrial ecosystems, so far as it goes, suggests that a rather sparse population of tracheophytes and some nematophytes occupied riparian and littoral sites and may have persisted through the dry seasons on the sites of ephemeral flood-basin ponds. Such marginally terrestrial ecosystems probably included some myriapods as detritivores and some of the amphibious scorpions and eurypterids as predatory foragers from adjacent aquatic habitats. Collembolids, mites, and mitelike forms occur in slightly younger assemblages and probably functioned as meiofaunal detritivores, herbivores, and predators; their absence from the Ludlovian-Pridolian is very likely an accident of sampling because such fossils are typically very rare even in post-Devonian deposits.

In summary, the fossil record shows clearly that brackish and freshwater ecosystems were well developed by the late Silurian. Although they may not have been much different from mid-Silurian ones, they are quite certainly different from those of the middle and late Ordovician. Brackish-water ecosystems apparently had relatively dense and diverse populations, including large detritivores and herbivores (agnathans and limulids) at the second and third trophic level and probably some fourth-level predators (acanthodians and eurypterids). Freshwater ecosystems, though less well represented, show similar patterns. Macrophytes had begun their deployment in terrestrial ecosystems, presumably with some effect on the micro- and meiofauna, but these forms are so inadequately represented by fossils that detailed inferences are impossible. However, these terrestrial macrophytes probably contributed to changes in aquatic ecosystems, through increases in nutrient release and in supply of organic detritus; they may also have reduced physical disturbance and instability in both aquatic and nonaquatic habitats by their effect on runoff, water storage, and erosion.

Gedinnian-Emsian

Marked changes occur in early Devonian continental assemblages, particularly those from freshwater and terrestrial habitats. Although some of this transformation may reflect more complete sampling, much of it represents real evolutionary innovation, particularly among tracheophytes and fish. For example, trilete spores are likely to be widely dispersed into marine as well as continental deposits and thus be least affected by sampling bias. They show a progressive increase in morphologic complexity and double in generic diversity from Pridolian to Gedinnian, double again in the Siegenian, and still again in the Emsian (Chaloner, 1972).

Late Silurian coastal lowland plant associations continued into the early Devonian but with some increases in morphological complexity, diversity, and size (Chaloner and Sheerin, 1979; Niklas, Tiffney, and Knoll, 1980; Knoll et al., 1984). In addition, lowland sediments yield some spores that cannot be attributed to the associated plant megafossils, thus implying transport from upland sites (K. C. Allen, 1981) and the appearance of an "upland" tracheophyte flora.

Radical changes also appear in the faunal record with the first appearance of a considerable variety of small arthropods (representing both nonaquatic and aquatic biocoenoses) and with the addition of a large number of new kinds of jawed fishes to the brackish and freshwater assemblages that continued from the late Silurian. As noted above, many (or all) of these small arthropods may have been present in middle Silurian (or even pre-Silurian) continental biocoenoses, though they have not yet been discovered as fossils. The timing of their first appearance may therefore have little ecologic or evolutionary significance. On the other hand, the appearance of the new groups of jawed fish must demonstrate a major change in aquatic ecosystems, for fossils of such fish would have almost certainly been found in late Silurian deposits if they had been present in ecologically significant numbers.

The characteristic Silurian brackish-water assemblages (of lingulids, ostracods, eurypterids, limulids, scorpions, anaspids, osteostracians, acanthodians) are modified by the occurrence of three major new taxa of jawed fish, the arthrodires, the crossopterygians, and the dipnoans, and by the further diversification of the jawless heterostracians (Størmer, 1976). The arthrodires and crossoptarygians represent predators that presumably fed on small fish and large invertebrates (discussed below). Early dipnoans (lungfish) also appear in some transitional marine-brackish situations, but not in fresh water, probably as predators on worms and arthropods (Denison, 1956, 429–30).

Freshwater assemblages of early Devonian age include a considerable variety of small crustaceans in addition to ostracods, fish, and (possibly) limulids. The crustacean fauna is dominated by brachiopods, including clam shrimps (Conchostraca) and fairy shrimps (Anostraca), as well as two orders, Lipostraca and Acercostraca, known only as fossils (Tasch, 1959, 140–41, 151–53, 159, 183–85). Among the jawless fish, the osteostracians (cephalaspids, ateleaspids, and two new groups, the kieraspids and galeaspids) continued in their benthic, mud-grubbing habitus; the freshwater heterostracians increased in abundance and diversity (for example, Halstead, 1973, 320–24), with new specializations for different feeding sites and different kinds of food items (Denison, 1956, 418–20; Halstead, 1973). Three orders of jawed fish representing at least two distinct classes — the acanthodians, the arthrodires, and the crossopterygians — are known from freshwater assemblages. The acanthodians are more diverse and abundant in early Devonian than in Silurian assemblages (increasing from one to five families) and, although primarily predatory on the small animals of the meiofauna, must have consumed some relatively large invertebrates and fish (Allen and Tarlo, 1963; Denison, 1979). The arthrodires make their first appearance in freshwater assemblages in the Gedinnian and invade brackish ones only in the Siegenian. The earliest species were probably limited to relatively small prey, such as crustaceans and fingerling fish, but later ones, at least as adults, fed on rather large animals (Denison, 1979). The crossopterygians, represented by only a few genera, also probably fed primarily on other fish (Denison, 1956, 428–30).

Early Devonian terrestrial faunas include myriapods (some possibly amphibious in habit; Størmer, 1976), collembolids, mitelike trigonotarbids, true mites (acarids), and spiders (Størmer, 1970, 1976; Scourfield, 1940; and summaries in Kevan, Chaloner, and Savile, 1975; and Rolfe, 1980). Generally these are types of animals that now inhabit soil and plant litter, some of them consuming algae, fungi, and bacteria, and others nematodes and other small arthropods (Rolfe, 1980). Though Kevan, Chaloner, and Savile (1975) suggest that some or many of these Devonian organisms fed on plant sap and spores, Rolfe (1980) questions their conclusion. In addition, some limulids, eurypterids, and scorpions may have foraged in terrestrial habitats (Størmer, 1976).

The early Devonian deposits of southern Britain have provided the most detailed information on specific continental habitats in the range from "proximal" (piedmont) and medial alluvial facies to the distal alluvial (lower delta plain) and coastal ones (J. R. L. Allen, 1979). The upland border or proximal facies consist primarily of playa, mudflow, and low-sinuosity channel deposits representing flashly hydrologic regimes; since they yield no plant fossils and only a few animal burrows, it appears that most aquatic and riparian macrophytes (and their accompanying consumers) were not yet able to invade such low-resource/high-stress habitats (Edwards, 1980). The medial alluvial facies also derive from low-sinuosity streams but seem to represent more stable habitats. In addition to drifted plant fragments they include dense, monospecific plant accumulations that suggest local stands of these plants on stream banks, upper portions of bars, and/or backwater swales (Edwards, 1980). Animal fossils include some fish remains, which are generally disarticulated and water-sorted, and an abundance and variety of trace fossils. No arthropods are reported, but the depositional situation was a poor one for preservation of small and/or delicate skeletons. The distal alluvial facies represents deposition by sinuous streams. Channel and channel-margin deposits yield drifted plant remains and many fish, including some pockets of articulated individuals associated with abundant plants in what may be swale deposits. In contrast, the overbank mudstones (nominally flood-basin swamp and pond habitats) yield only trace fossils as the sole remnants of an autochthonous biota and scattered, drifted plant remains.

The transformations in the composition of continental biocoenoses and in the characteristics of continental ecosystems in the early Devonian are parallel to, but much more extensive than, those noted for the late Silurian. The most significant changes in composition (so far as ecosystem development is concerned) include (1) increase in abundance, diversity, and morphological complexity of semiaquatic and terrestrial plants, primarily tracheophytes; (2) increase in variety and abundance of aquatic and terrestrial meiofauna, primarily arthropods; and (3) increase in abundance and variety of aquatic macrofauna, primarily heterostracan and gnathostome fishes. These presumably reflect increases in primary production, reductions in disturbance and instability, and alterations in trophic structure and dynamics.

The changes in the continental floras were almost certainly basic to those in consumer populations and to the major alterations in the ecosystems. The increased abundance and wider occurrence of macrophytes clearly demonstrate that primary production was greater in the early Devonian than in the late Silurian, although primarily limited to riparian and lake-margin situations in alluviodeltaic lowlands. The rhizome system of early tracheophytes (particularly if associated with mycorrhizae and mycotrophism, as suggested by Pirozynski and Mallock, 1975, and as implied by some fossil material from the Rhynie Chert of Scotland) and the chemical environment created by decomposition of plant litter would have increased release of nutrients from the substrate. This in turn would increase the potential concentration of nutrients in the surface water and in the associated aquatic and terrestrial biomass. These changes, together with the recycling of buried nutrients by rooted hydrophytes, would also have increased the potential nutrient supply for aquatic microphytes and hence their productivity. In addition, the extensive development of prostrate rhizomes with rhizoids must have stabilized terrestrial soils and aquatic sediments against rapid erosion (Edwards, 1980) and damped variation in soil water, surface runoff, and stream flow, at least in relatively well vegetated lowland areas. Such vegetational cover (and the associated plant litter) would also have provided some protection against water loss and temperature variations for small organisms as well as some stability in food supply. On the other hand, the character of the earliest Devonian tracheophytes — for example, their small size and the lack of leaves and true roots in many groups — suggests that primary productivity was still relatively low and moderation of physical disturbance and uncertainty was rather limited. These latter conclusions accord with the apparent limitation of tracheophytes to medial and distal alluvial facies and, probably, to riparian and channel swale situations.

Consumer populations in freshwater biocoenoses changed markedly, whereas similar changes in brackish-water and terrestrial consumers (and ecosystems) appear less ecologically significant. In freshwater ecosystems, the expansion of invertebrate consumers (at second and third trophic levels) is reflected not only in occasional body fossils, but more clearly in the abundant and varied trace fossils, even in overbank pond habitats. Fourth-level macropredators — such as acanthodians, arthrodires, and crossopterygians — were rare or nonexistent in earlier freshwater ecosystems but are relatively abundant and diverse in Devonian ones. Diversification within habitats at the second and third levels (heterostracians and some acanthodians and arthrodires) must also indicate significant changes in trophic structure and/or habitat characteristics. This general pattern is further confirmed by the appearance of a variety of small arthropods as primary consumers (though the total absence of comparable forms from earlier assemblages might be ascribed more to inadequate sampling than biologic reality). The increase in fossil abundances further suggests a significant increase in secondary production.

In constrast, the consumers in terrestrial ecosystems apparently consisted largely or entirely of small animals operating on the second trophic level (with a preponderance of detritivores over herbivores), with a modest variety of small predators on the third trophic level. Any comparison with late Silurian terrestrial consumers is dubious, however, since the absence of a meiofauna from that interval may be a function of limited and biased sampling. Although one might expect that some changes occurred in consumers consequent on expansion of terrestrial macrophyte populations, the characteristics of these early Devonian fossil representatives suggest their continued limitation by low or episodic primary production and by intense physical disturbance.

Brackish-water ecosystems, in spite of greater diversity at the third and fourth trophic levels and some large changes in taxonomic composition, show only rather minor changes in basic characteristics, for example, in trophic structure. Thus it appears that the diversification of brackish-water ecosystems was largely accomplished by the end of the Silurian — in contrast to that of the freshwater ones, in which diversification continued through the Devonian.

Eifelian-Famennian

The tracheophyte "green revolution" progressed through the middle and late Devonian with the development of rooted plants and the appearance of shrubs and trees in terrestrial habitats, and with the further enrichment of freshwater ecosystems. Known faunal changes were concentrated largely in brackish-water and freshwater fish; terrestrial faunas, so far as they are known, show relatively little modification in composition — late Devonian and early Carboniferous amphibians appear more closely tied to aquatic ecosystems than to terrestrial ones (Olson, 1971, 631, 634–36).

Among the tracheophytes, large, pinnate leaves appear in the Eifelian with the evolution of progymnosperms and pteropsids. Some representatives of these groups as well as of the lycopsids reached tree size and had well-developed root systems and foliage crowns (Chaloner and Sheerin, 1979). Both progymnosperms and lycopsids developed heterospory, with a very much enlarged megaspore providing a food reserve for the gametophyte and embryonic sporophyte — a feature that would be desirable where shade from a plant canopy limited seedling survival and growth (Chaloner and Sheerin, 1979). Development of foliage and increase in size required (or depended on) transformations in structure of the tracheids, reorganization of the xylem, and alteration in growth pattern (Banks, 1970, 97–90; Chaloner and Sheerin, 1979). Smaller plants persisted among the lycopsids and evolved in the pteropsids, though some primitive groups of small plants (for example, rhyniophytes) diminished in importance after the Eifelian (Chaloner and Sheerin, 1979). As a result, spatial stratification appeared in terrestrial communities (Niklas, Tiffney, and Knoll, 1980; Tiffney, 1981; Edwards, 1980; Chaloner and Sheerin, 1979), with possible implications for the emergence of winged insects in the Carboniferous.

In brackish and freshwater biocoenoses, the vertebrates underwent considerable change. Agnathans, both osteostracians and heterostracians, diminish progressively through the later Devonian (Westoll, 1979) with only a few persisting until the Frasnian. Some of the benthic, detritus-feeding niches were filled by the antiarchs, which derived from an arthrodire stock in Siegenian and are the commonest fish fossils in most middle and late Devonian freshwater deposits. Arthrodires continued as medium-size predators; acanthodians maintained a similar role. The osteichthians underwent the most marked changes — the appearance of the first ray-fins (actinopterygians), the invasion of fresh waters by lungfish (dipnoans), and a radiation of the crossopterygians (Westoll, 1979). The early actinopterygians were small- to medium-sized fish and were probably primarily predators on small nektonic and benthic animals, apparently filling heterostracian-like roles in part. The lungfish seem to have been dominantly selective bottom feeders, fitted to prey on tough-shelled crustaceans and on annelids (Denison, 1956, 429–30). The crossopterygians were medium- to large-sized nektonic predators that as adults must have fed on other fish and large invertebrates. In addition, some or all of the early amphibians (known from two Frasnian trackways and a few Famennian skeletons) probably fed primarily or entirely in aquatic habitats on similar types of prey (see discussion below).

Mid- and late Devonian freshwater and terrestrial arthropods are less well known than early Devonian ones — though, obviously, groups with both early Devonian and post-Devonian occurrences must have been present: for example, the fairy and clam shrimps among the crustaceans; the scorpions, acarids, trigonotarbids, and spiders among the archnids; the myriapods; and the collembolids. Kevan, Chalone, and Savile (1975) suggest that many of the modifications in structure and size in mid- and late Devonian plants represent adaptive responses to attack by an increasing number of herbivorous arachnids and hexapods. On the other hand, eurypterids seem to have declined in significance after the early Devonian; for example, twelve genera are reported from late Silurian and fifteen from early Devonian but only nine from middle and eight from late Devonian (Størmer, 1955).

Whether any of the early amphibians foraged in terrestrial ecosystems is uncertain; they may have been primarily transients on their way from one pond to another. Certainly the skeletons known from the very late Devonian, with their large size and fishlike teeth, jaws, and tail, suggest animals that, at least in their adult stage, fed primarily or entirely on fish and large aquatic invertebrates and traveled over land primarily to feed in scattered freshwater habitats (Olson, 1971, 635–36). On the other hand, juvenils of these species might have foraged for small terrestrial as well as aquatic invertebrates, and some species with small, arthropod-feeding adults may well have been present in riparian sites.

These changes in freshwater and terrestrial biocoenoses imply significant transformations in ecosystem structure and dynamics. In particular, the appearance of tree-shrub communities (open forest?), at least in well-watered coastal lowlands, implies an enlarged food supply for terrestrial herbivores and an increased export of organic detritus as well as dissolved nutrients to aquatic ecosystems. In addition, their presence would have further moderated physical disturbance and instability in aquatic as well as terrestrial habitats, although the total effect may have still been relatively small because of the limited geographic extent of such forests. Finally, to the extent that tracheophytes, particularly trees, spread into the continental interiors, the increased return of moisture to the atmosphere through transpiration and the reduction in terrestrial albedo in the tropics may have broadened the extent of tropic, wet climates and diminished that of subtropical dry climates (see discussion by Charney, Stone, and Quirk, 1975).

The transformations in freshwater ecosystems, presumably induced by their coupling with terrestrial systems, are striking. They include not only continuation of earlier trends in trophic differentiation and specialization but also the appearance of new trophic tactics, that is, foraging among scattered, isolated aquatic habitats by amphibious vertebrates. At the second and third trophic levels the agnathans were replaced (or displaced) by antiarchs, arthrodires, and actinopterygians, but the available evidence is not adequate to demonstrate any other changes at this level — for example, changes in productivity or in number, type, or breadth of niches. However, the radiation of crossopterygians suggests significant changes at the third and fourth levels. This may have involved one or more of the following: (1) increased supplies of previously rare food items, (2) utilization of previously inhospitable habitats, and (3) splitting and narrowing of niches. If, as suggested above, the early amphibians fed primarily by foraging among small, more or less isolated aquatic ecosystems, they would effectively have added another trophic level (the fourth) to many of those systems as well as competing with crossopterygians for niches in other aquatic habitats.

The effects of eutrophication also appear in the assemblages representing large lakes. For example, Donovan (1980) ascribes the abundant and diverse fish faunas of the middle Devonian of Caithness in Scotland to a deep, eutrophic tropical lake supporting three or four link food chains based on primary production by pelagic microorganisms. The absence or at least rarity of such ecosystems before the middle Devonian may be attributed to a low input of detritus and dissolved nutrients prior to development of an extensive cover of land tracheophytes, together with their bacterial and fungal consorts.

Changes in terrestrial ecosystems are obscured by the rarity of terrestrial arthropods in middle and late Devonian fossil assemblages and by the very limited evidence for amphibian evolution and ecology. As already noted, Kevan, Chaloner, and Savile (1975) interpret some features of the plants as indicative of increased consumption by arthropods, and thus of the presence of such forms in the biocoenoses if not in the fossil record. In particular, they suggest that the diversification of the hexapods and the initial stages in evolution of winged insects were coupled with changes in the flora. (Recent discoveries of arthropods in the late Devonian of New York seem to support some of these suppositions; Shear et al., 1984). As suggested above, some of the early amphibians may have utilized terrestrial arthropods as a major food source at the third and fourth trophic levels. The fossil record gives no positive evidence for this inference, however, and in any case the extreme rarity of tetrapods in late Devonian assemblages is presumptive evidence against a major role for macroconsumers (at the fourth trophic level) in terrestrial ecosystems.

Overall, late Devonian continental ecosystems were characterized by further differentiation and expansion of consumers at the third and fourth trophic levels, although this is better documented for freshwater than for brackish or terrestrial habitats. This change was associated with a very marked diversification of terrestrial tracheophytes, along with an increase in their abundance and an expansion in their habitat range. These transformations presumably increased primary production in terrestrial habitats and the supply of dissolved nutrients and organic detritus to aquatic ones, but the rarity of large accumulations of plant debris (preserved as coals) demonstrates that productivity had yet to reach the level attained in the Carboniferous and later. Changes in secondary production are more difficult to demonstrate, but in aquatic assemblages the increased abundance and diversity of animals at the top trophic level argues for higher production at intermediate levels. This conclusion presumably is supported by concentration in aquatic habitats of increased plant detritus from terrestrial habitats. Conversely, the apparent absence, or near absence, of macropredators from terrestrial ecosystems until the mid-Carboniferous points toward low secondary production or a temporal lag in evolution of terrestrial consumers.

Patterns in Development

Directions of Change

The geologic record demonstrates an increase in the number and variety of fossils and fossil assemblages representing continental ecosystems from the Ordovician through the Devonian; it also demonstrates changes in associated sedimentary rocks that reveal biological effects on weathering, erosion, and deposition. Analysis suggests that these changes involved large-scale transformations in (1) trophic structure and dynamics, (2) habitat utilization, and (3) ecological strategies of brackish, freshwater, and terrestrial ecosystems. These inferences are documented primarily by macrofloral and macrofaunal fossils but are also consistent with the limited number of meiofaunal samples and with changes in depositional style related to abundance and distribution of plants. In addition, they are compatible with deductions about early continental habitats drawn from the absence or extreme rarity of fossils.

At their first appearance, assemblages in all three types of continental habitat are very rare, consist of relatively few fossils, include few if any forms that might have fed at the fourth trophic level, and show little variety even at the third level. Deposit/detritus feeders (at the second and third trophic levels) predominate although they display little diversity. In addition, the sediments show little evidence of biologic activity — for example, the accumulation of reduced carbon, changes in the intensity and mode of chemical weathering, changes in bioturbation or in the patterns of alluvial deposition. Overall, trophic structure, at least as represented by macrofossils, was relatively simple; there was (1) little differentiation between second, third, and fourth levels, (2) little diversification and specialization within levels, and (3) a dominance of detritus as a food source.

Brackish-water assemblages display these "primitive" characteristics at their first appearances in the Ordovician but undergo progressive changes in the Silurian and Devonian; such assemblages are rare through the middle Silurian but are common thereafter. Although evidence for early and middle Silurian biocoenoses is scanty, the fourth trophic level was clearly differentiated by this time, and there is some suggestion of diversification at levels two and three. However, deposit feeders remained the dominant trophic type, although some meio- and macropredators were present. Trophic diversification was continued in late Silurian and early Devonian ecosystems with an increase in the importance of predators relative to deposit feeders and with specialization for different kinds of prey and for different feeding sites at each trophic level.

Freshwater ecosystems displayed a similar pattern of development, although lagging behind the brackish-water ones. The earliest assemblages (early and middle Silurian) are rare and represent relatively simple trophic structures. Occurrences in the late Silurian are more common and more fossiliferous, but show that trophic structure was apparently similar to that of mid-Silurian: (1) little diversity or specialization at the second and third trophic levels, (2) little or no differentiation at the fourth level, and (3) dominance of deposit feeders. By the end of the early Devonian, however, the situation had changed: the consumers on the second and third trophic levels had diversified and specialized further by food type and feeding site, and the fourth level was represented by several types of fish and, possibly, by macropredatory arthropods. Diversification and specialization continued into the middle and late Devonian, culminating with the appearance of large macropredators that foraged among separate aquatic habitats as well as within them. In this event, the trophic structure (if not trophic diversity) approximated that present in Carboniferous and younger freshwater ecosystems.

Development of terrestrial ecosystems followed a step or two behind the aquatic. Early and middle Devonian assemblages were probably dominated by small detritus feeders, although some meiopredators (effectively third level) were present. Sampling of those components is so spotty, however, that generalizations about diversification and specialization are impossible. The appearance of the first tetrapods in late Devonian time potentially added a fourth trophic level to terrestrial ecosystems, but the extent of their participation in terrestrial, as opposed to aquatic, food chains is doubtful. Indeed, for terrestrial ecosystems extensive development of the fourth trophic level and diversification and specialization at the second and third is not clearly documented prior to the late Carboniferous.

Primary and secondary production apparently increased in all three habitat regimes in parallel with, or perhaps in advance of, changes in trophic structure. Though the fossil record provides no direct, unambiguous evidence of productivity, inferences from fossil abundance and other biogenic inputs to deposition do provide a semiquantitative measure of change. Where assemblages with similar preservational potential can be compared, relative fossil abundance bears some relation to production. On the basis of such evidence, secondary production increased through the Silurian into the Devonian. In brackish-water ecosystems the change seems most marked from mid-Silurian to early Devonian and may have approached a plateau thereafter. In freshwater systems, abundance rose sharply after the late Silurian/earliest Devonian and continued an upward trend thereafter — perhaps not culminating until the late Carboniferous/early Permian. The increase in macrophyte remains in aquatic assemblages through the Silurian and Devonian probably records an increase both of *in situ* primary production and of import of allochthonous plant detritus; in either case, it marks an increase in energy available for consumers. In terrestrial ecosystems the rapid increase in abundance of terrestrial plant fossils from late Silurian through late Devonian indicates increased primary production. This conclusion is supported by the shift in the dominant mode of alluvial deposition from braided- to sinuous-channel deposits reflecting the influence of the plants on stream discharge and sediment bed and on bank cohesiveness. Increased primary production also implies an increase in secondary production, but fossil evidence for consumer abundance is insufficient to test this inference.

These trends in ecosystem characteristics were apparently associated with major changes in the physical characteristics of continental habitats and in utilization of these habitats. I argued above that pre-Silurian continental habitats

1. were marked by very low availability of resources (nutrients and water),
2. presented very high levels of physical disturbance affecting resource utilization, and
3. were subject to frequent, intense, and rather unpredictable variations in the physical environment.

These constraints on biological activity and organization would have been least significant in near-marine habitats and most severe in terrestrial ones. The inferred increases in trophic complexity and in production during the later Silurian and the Devonian clearly indicate relaxation of some or all of these constraints. The features of Silurian and Devonian tracheophytes demonstrate that some of the change was consequent on new adaptations for acquisition and conservation of scarce resources and for resistance to physical disturbance, but the reduction in physical constraints also involved biogenic modifications of the physical habitats themselves.

For example, in terrestrial habitats the evolution of rhizoid-rhizome and root systems in tracheophytes (and the development of the ectomycorrhizal consortium) must have not only increased nutrients for the individual plants themselves but also augmented and conserved the general stock of dissolved and of organically bound nutrients. Further, the appearance of an herbaceous and later an arborescent plant cover with an associated litter-and-root zone must have reduced water loss, increased humidity near the ground, limited temperature variation, and reduced erosion. In addition to increasing resource availability still further, these modifications reduced desiccation and thermal stress and permitted more effective utilization of soil nutrients by plants and consumption of plant tissues and detritus by animals. Obviously, such environmental amelioration wrought by the appearance of terrestrial plants would have produced a positive feedback; that is, the increase in resources and reduction in disturbance and instability would have facilitated additional plant growth and thus furthered modification of physical circumstances. The overall consequence was a progressive increase in what one might call "ecosystem homeostasis," complementing individual homeostatic capacities.

Parallel changes must also have occurred in aquatic habitats as a consequence of the development of a terrestrial (and riparian) plant cover and the associated litter-root-soil complex. The storage of water in the soils and the high humidity under the plant canopy would have moderated fluctuations in runoff and thus in stream discharge. Reduced erosion of slopes, gullies, and banks (owing to the presence of rhizome-rhizoid and root systems) would have reduced the amount of coarse sediment supplied to streams. The result would have been more rivers with more equitable discharge and more stable channels and bars. These effects would also lead to a stabilization of water level in flood-basin lakes and ponds and to a moderation of fluctuations of turbidity and salinity in brackish estuaries.

These deductions are supported in part by sedimentologic and paleobiologic evidence. As noted above, stream deposition tends to change in character from braided- to sinuous-channel regimes, marking a shift toward less episodic runoff, erosion, and deposition and toward a retention of finer-grained sediment. An increase in the abundance of swamp deposits from the mid-Devonian into the Carboniferous also suggests a change from "flashy" to "prolonged" floods. Further, some of the increase in diversity within assemblages may be due to reduction in disturbance; Grime (1979) suggests that diversity is typically low in habitats subject to frequent, intense disturbance (and/or to low resource availability). The progressive extension of ecosystem distribution during the Devonian into inherently more rigorous (and resource-poor) habitats — for example, those of piedmont alluvial plain

— probably also reflects some reduction in physical stress and disturbance.

Factors in Ecosystem Development

What controlled these patterns of development, namely, the elaborations of trophic structure, the increases in productivity, and the diachrony in phases among habitat types? Moreover, why do complex ecosystems appear later in continental habitats than in marine? Finally, why were so many of the critical changes concentrated in an 80-million-year period (late Ordovician to late Devonian) but still widely dispersed within that interval?

Ecological Constraints

Some of the answers must lie in the relationships of ecosystem structure and dynamics to the general characteristics of continental habitats. Compared with marine habitats, the early continental habitats provided little in the way of readily accessible resources and exposed their occupants to severe physical stress and high spatial and temporal uncertainty. As noted above, resource shortages must have limited primary production and favored resource conservation above expenditures for maintenance, growth, and reproduction. In turn, low primary production would restrict secondary production and favor an emphasis on resource conservation among consumers as well as producers. On the other hand, physical stress and uncertainty would induce low average survivorship and favor (or require) an emphasis on replacement (growth and reproduction); such emphasis runs contrary to an emphasis on conservation. With low resources for production and with high requirements for replacement, only a small number of individuals could maintain themselves; consumer populations were limited to the lower trophic levels (where resources were most abundant) and to the most readily available type of food at each level; consumers would also tend to feed as generalists within and among levels. In addition, unpredictable variations in physical factors would produce marked fluctuations and tend to destablize food chains, particularly long ones where disruption would have been multiplicative (see Pimm and Lawton, 1977). The overall result in ecological time was to limit the number of trophic levels and the number of potential niches at each level. (The emphasis on conservation and/or replacement also had consequences in evolutionary times, as described in the following section.)

Some of the adaptive innovations that appeared in the evolution of continental macrophytes increased their access to resources and reduced the effects of physical stress and instability on resource utilization and on survival. This led to a slow shift in emphasis in adaptation from conservation and replacement to resource acquisition (or to competitive preemption) and to maintenance; this shift led to the appearance of a plant canopy, a thick, widespread litter zone, and an equally extensive development of subterranean rhizomes, rhizoids, roots, and ectomycorrhizae. These features increased "ecosystem homeostasis" through conservation of water and nutrients, through reduction of temperature extremes, and through modification of erosion and deposition. The increase in net primary production allowed an increase in secondary production; cumulative increases in primary and secondary production added to the supply of each kind of potential food item as well as to the total number of kinds of food. The moderation of physical stress and uncertainty reduced requirements for reproduction; it also reduced the fluctuations in population densities and the consequent destabilization of food chains. Overall, the result in ecologic time was an increase in the potential number of trophic levels and of niches at each level. In evolutionary time, these changes favored increased specialization of organisms and niche splitting. This led, in ecologic terms, to increased trophic differentiation, specialization, and diversification.

What of the differences among brackish, freshwater and terrestrial ecosystems in the timing of ecosystem development? The rather moderate levels of physical stress and of uncertainty, and the high influx of nutrients and water from a marine reservoir, account for the precocious emergence of advanced, complex brackish-water ecosystems. Freshwater ecosystems, which were subject to more physical stress and disturbance than brackish ones and which were much more dependent on release of nutrients from terrestrial habitats, were somewhat slower to develop. On the face of it, the belated emergence of the upper trophic levels in terrestrial ecosystems (really not until mid-Carboniferous) could be ascribed to the high intensity of habitat stress and uncertainty, rather than to resource supply, which should be roughly equivalent in freshwater and terrestrial habitats. On the other hand, a gross increase in biomass of terrestrial vascular plants does not insure an increase in assimilation by terrestrial consumers; much of this biomass (and much of net primary production) must have consisted of tissues of low nutritive value (high fiber/low protein), compared with the decomposing detritus and the primary production available to aquatic consumers. Thus, the evolution of terrestrial herbivores might have been hindered and delayed by these and other unfavorable characteristics of vascular plants as food sources. Therefore the middle and late Devonian increases in terrestrial primary production may have been of more value to aquatic than terrestrial consumers and may not have provided much basis for increases in secondary production among the latter.

Evolutionary Aspects

If innovations among continental plants for resource acquisition and for maintenance were the keys to the development of continental ecosystems, what factors controlled *their* evolution? Further, what specific conditions or events determined the mode and timing of their appearance?

I suggested above that the severe resource shortages and the intense physical disturbance characteristic of early continental habitats placed particular emphasis on adaptations for resource conservation and/or for replacement (reproduction and recolonization) relative to those for resource acquisition and resistance to disturbance. In modern resource-poor situations, organisms typically conserve whatever resources are available through very slow growth with very limited investments in (1) resource acquisition, for example, in growth of collecting systems for light, nutrients, and water; (2) resistance to disturbance; and (3) reproduction (Grime, 1977, and 1979, 33–36). This conforms to the classical concept of a "K strategy," of which the lichen consortium seems to be the ultimate example — adaptation to extreme resource shortages (Grime, 1979, 36). In contrast, where circumstances demand high replacement rates, reproductive capacity tends to be emphasized over resource acquisition or resistance to disturbance. The result is dominance by classical "r strategists," forms marked by small size, rapid growth, short lives, simple structure, and capacity for wide dispersal (Grime, 1977, and 1979, 39–45). Many soil and freshwater microphytes represent an extreme version of this adaptive pattern. In intermediate situations, intermediate characteristics appear, basically such that resource conservation mechanisms are imposed on (and limit) replacement capacity. Grime (1979, 65–66) argues that some mosses and liverworts are exemplars of this pattern, near the extremes both of resource limitations and of disturbance. The ecologic distribution of modern tracheophytes suggests that, even with their homeostatic adaptations for resource acquisition and maintenance, they are unable to tolerate extremely low resource levels or very severe disturbance; habitats marked by these features, or by combinations of them, are left largely to lichens and microphytic algae.

This situation implies that the predominant selective forces operating in such extreme habitats oppose adaptations that would require a high initial investment of resources and, by extension, that such "counterselections" in early continental habitats probably opposed evolution of the adaptations for resource acquisition and for regulation of disturbance that characterize most continental macrophytes. Thus the generation and survival of continental macrophyte lineages becomes relatively unlikely and their establishment a rather improbable kind of event, one that was, at best, only weakly predictable (or determined) from the ecological setting because of counterselection. The hypothesis that counterselection limited the growth of lineages would be falsified if (1) adaptive patterns in early macrophytes were dominantly for resource acquisition rather than for conservation and/or reproduction; (2) several phylogenetically distinct groups of macrophytes originated independently during the same short time interval; (3) macrophytes originated in resource-poor and/or intensely disturbed habitats; or (4) the pattern of macrophyte evolution was generally the same after establishment of those homeostatic adaptations that increased access to resources and ameliorated environmental disturbance as it was before. All four predictions can be tested on the basis of fossil evidence. If the hypothesis survives such tests, it then provides a base-level, null hypothesis with which to evaluate other scenarios for macrophyte evolution.

The hypothesis appears to pass the first test although characterization of adaptive patterns in earliest continental macrophytes must be somewhat tentative. Of the earliest occurrences, the small, thick-walled spores suggest an adaptation for survival and dispersal in severely disturbed circumstances (Grime, 1979, 115). The cuticlelike sheets of cells suggest adaptation for conservation of water in the face of severe desiccation. Certainly, the very restricted development of both aerial and subterranean components in Silurian and even early Devonian tracheophytes is consistent with a resource conservation pattern (Grime, 1977), and the emphasis among early tracheophytes on vegetative reproduction (Tiffney, 1981; Tiffney and Knoll, in press) is also compatible with survival in resource-poor situations (Grime, 1979, 113–14). Even the erect form may be as related to spore dispersal (Tiffney, 1981) as to the acquisition of light. This is not to argue that the features are entirely the consequence of adaptations for resource conservation. The lack of roots might be due to the temporary lack of a developmental solution to the problem of anchorage and of nutrient and water uptake; the occurrence of prostrate rhizomes might be the result of structural weaknesses in shoot tissues, rather than primarily adaptive to other functions incidental to an obligate prostrate form. Rather, none of these features indicate extensive investment in resource acquisition and therefore do not falsify the hypothesis. Only the vascular system and the stomates can arguably be related to increased acquisition of resources and increased photosynthetic activity. The nematophytes are less well understood than the tracheophytes, but what is known suggests an adaptive pattern comparable to that of tracheophytes, since nematophytes lack stomates for carbon dioxide intake and have yet to yield any definite traces of a rhizome-rhizoid system.

On the other hand, the rough coincidence in time of appearance of nematophytes and tracheophytes (within 20 million years or less) appears to negate the hypothesis. If the events were truly independent and could have occurred randomly at any time after the appearance of continental eukaryotric algae, perhaps 800 million years ago, their occurrence by chance in the same 20-million-year interval would seem to be very small. However, the critical assumptions in this argument are that the events are "truly independent" and that they could have occurred "randomly at any time." The former assumption almost certainly does not hold, even if nematophytes and tracheophytes had completely separate origins from terrestrial or aquatic microphytes. The appearance of either group, even in prototypic form, would likely have increased local resource levels and/or moderated disturbance, thereby creating richer and less hostile habitats in which the intensity of counterselection would have been relaxed and the likelihood of homeostatic adaptations increased. In effect, the evolution of one group could have induced the origin of others.

The second condition, temporal randomness, is also unlikely. Certainly, geographic and climatic fluctuations through the late Proterozoic and early Paleozoic must have produced major changes in the number of habitats favorable for macrophyte evolution and thus induced some degree of temporal determinism. Therefore the null deduction must be restated: the counterselection hypothesis is incorrect (as stated) if the number of relatively rich, undisturbed habitats was exceptionally small from the late Ordovician through the early Silurian. Here the evidence — and interpretations of it — are quite ambiguous. Pratt, Phillips, and Dennison (1978) argue that continental macrophytes arose from preexisting marine macrophytes when the latter were exposed to increased disturbance (and presumably decreased resources) during low stands of sea level in the late Ordovician and early Silurian. Gray and Boucot (1978; also Boucot and Gray, 1982) deny such an association; they point out that changes in sea level would have been sufficiently slow to permit migration of marine populations as the geographic distribution of neritic habitats changed. I add that there is no strong evidence that either nematophytes or tracheophytes originated from marine macrophytes (Stebbins and Hill, 1980) or that fluctuating sea level would produce a decrease (or increase) in the number of habitats favorable for evolution of either lineage from aquatic or terrestrial microphytes. Gray and Boucot (1977) suggest a similar linkage to the appearance of monsoonal climatic regimes in the Silurian and Devonian. The mechanism would again be intensified selection associated with seasonal desiccation of alluviodeltaic environments; however, as Gray and Boucot observe, this would require that pre-Silurian climates lacked monsoonal or other wet/dry regimes — a condition for which no evidence is available at this moment. Overall, the counterselection hypothesis is not falsified by evidence now available on the abundance of favorable habitats in the critical Ordovician-Silurian interval — but then, neither is it supported.

The third potential test, the character of specific habitats in which macrophytes originated, also seems to be met by the counterselection hypothesis, so far as interpretation of those habitats can be pushed. The early macrophyte occurrences represent either lowland aquatic or riparian alluviodeltaic habitats (or assemblages transported from such situations into adjacent, near-shore marine environments); these are situations that would have been relatively rich in resources and little disturbed. In addition, the spore assemblages from these rocks give no evidence for any floral sources beyond the recorded lower coastal plain assemblages (K. C. Allen, 1981). Finally, even as late as the middle and late Devonian, tracheophyte remains are very rare or totally absent in "upper" alluvial plain deposits and even in nonriparian lowland sites. Again, such evidence does not "prove" the hypothesis but is at least consistent with it.

The results of the fourth test, the patterns of evolution before and after establishment of macrophyte-level maintenance and resource access, also conform to the hypothesis rather than falsify it. A radical change in macrophyte evolution does occur from the late Silurian onward. The evolutionary mode appears more strongly determined, more predictable, and primarily for features that augment or complement resource acquisition and, for the first time, resource preemption; they also show far less emphasis on reproductive capacity and more on maintenance. Thus in 30 million years or less, most of the principle macrophyte adaptive types were well established. Increased capacity for maintenance and resource acquisition reduced counterselective pressures for small size, limited growth, and simple structure, and reinforced selection for more efficient maintenance and resource acquisition.

Although a nondeterministic hypothesis for the *timing* of macrophyte evolution cannot be falsified on present evidence, that hypothesis (or any nondeterministic one) provides a parsimonious baseline (null hypothesis) for evaluation of deterministic alternatives. Such an alternative must not only withstand falsification but must be preferable in some other way to a nondeterministic one. In particular it must explain features of the record not otherwise accounted for or, at least, be demonstrably necessary as well as sufficient to explain the timing as well as the pattern of macrophyte evolution. These are quite restrictive conditions, but they seem essential if one is to reduce the "just-so" elements in explanation of historical phenomena. These criteria leave us, in my judgement, without a "favored hypothesis" to explain the timing of macrophyte evolution in continental communities.

Conclusions

Early development of continental ecosystems seems related primarily to a handful of adaptations critical to early macrophyte evolution. These initially involved establishment of effective mechanisms for survival of physical disturbance and for conservation of essential resources. These events most likely occurred in relatively rich and undisturbed habitats where sufficient time and resources were available to support such relatively costly maintenance devices. They were followed by the appearance of features that increased acquisition, utilization, and ultimately preemption of resources, such as taller axes, foliage, and roots. In sum, these adaptations resulted in an increase in primary production and a moderation of disturbance in continental habitats. The ecosystem response, in evolutionary time, was an increase in secondary production and an elaboration of trophic structure marked by increased differentiation, specialization, and diversification at the third and fourth trophic levels. Ecosystem changes began first in habitats in which resources were readily available and disturbance relatively low (that is, in brackish habitats) and proceeded more slowly in freshwater habitats, where resources were less readily available and disturbance more intense, and still more slowly in terrestrial habitats. The pattern of change suggests that resource availability may have been the constraining factor for most aquatic ecosystems and physical stress the constraint for most terrestrial ones.

The early phases of macrophyte evolution were, at the most, weakly predictable because of strong counterselection against energetically expensive adaptations in characteristically resource-poor, intensely disturbed continental habitats. However, accumulation of such adaptations in some of those relatively rare habitats where conditions were more favorable switched the later phase of macrophyte evolution to a strongly predictable one dominated by resource acquisition as counterselection for resource conservation and/or mortality replacement relaxed. Neither evidence nor theory support a converse hypothesis. This model provides a sufficient hypothesis for the pattern of early macrophyte evolution and, since it involves a high level of indeterminacy, provides a basic "null hypothesis" for interpreting those events. Any alternative hypothesis must be shown to be not only sufficient but necessary if it is to be favored over such a null one; such necessity remains to be demonstrated.

Acknowledgments

The general approach of this paper was inspired by the work of Everett C. Olson; Douglas Grierson not only assisted with the paleobotanical aspects but provided dogged but vital criticism of my ecological and evolutionary extravagances as these ideas developed over the past six years. I also profited from a discussion with Ian Rolfe of the more esoteric aspects of the record of early terrestrial invertebrates. He and Everett Olson provided helpful reviews of an earlier draft of the paper; the present version was improved by the many thoughtful suggestions (as well as pointed criticisms and close editing) of Bruce Tiffney. I retain full responsibility for such deficiencies as have escaped their eyes or their patience.

References

Allen, J. R. L. 1979. Old Red Sandstone facies in external basins, with particular reference to southern Britain. In *The Devonian system*, ed. M. House, 65–80. London: Palaeonotological Association.

Allen, J. R. L., and L. B. Tarlo. 1963. The Downtonian and Dittonian facies of the Welsh borderland. *Geol. Mag.* 100:129–55.

Allen, K. C. 1981. A comparison of the structure and sculpture of *in situ* and dispersed Silurian and early Devonian spores. *Rev. Palaeobot. Palynol.* 34:1–10.

Banks, H. P. 1970. *Evolution and plants of the past.* Belmont, CA: Wadsworth.

———. 1972. The stratigraphic occurrence of early land plants. *Palaeontology* 15:365–77.

———. 1975a. Early vascular land plants: Proof and conjecture. *Bioscience* 25:730–37.

———. 1975b. The oldest vascular plants: A note of caution. *Rev. Palaeobot. Palynol.* 20:13–25.

Basu, A. 1981. Weathering before the advent of land plants: Evidence from unaltered detrital k-feldspars in Cambrian-Ordovician arenites. *Geology* 9:132–33.

Boucot, A. J., and J. Gray. 1982. Geologic correlates of early land plant evolution. *Proc. Third North American Paleontological Conf.* 1:61–66.

Boucot, A. J., and C. Janis. 1983. Environment of the early Paleozoic vertebrates. *Palaeogeogr., Palaeoclimatol., Palaeoecol.* 41:251–87.

Boureau, E. A., A. Lejal-Nicol, and D. Massa. 1978. A propos du Silurien la date d'apparition des plantes vasculaires. *Acad. Sci., Paris, C. R., Ser. D* 28b:1567–71.

Brown, D. A., K. S. W. Campbell, and K. A. W. Cook. 1968. *The geologic evolution of Australia and New Zealand.* New York: Pergamon Press.

Caster, K. E., and E. N. Kjellesvig-Waering. 1964. Upper Ordovician euripter-ids of Ohio. *Palaeont. Americana* 4:297–358.

Cawley, J. L., R. C. Burruss, and H. D. Holland. 1969. Chemical weathering in central Iceland: An analog of pre-Silurian weathering. *Science* 165:391–93.

Chaloner, W. G. 1972. The rise of the first land plants. *Biol. Rev. Cambridge Philos. Soc.* 45:353–77.

Chaloner, W. G., and A. Sheerin. 1979. Devonian macrofloras. In *The Devonian system,* ed. M. House, 145–61. London: Palaeontological Asso-ciation.

Charney, J., P. H. Stone, and W. J. Quirk. 1975. Drought in the Sahara: A biogeophysical feedback mechanism. *Science* 187:434–35.

Cotter, E. 1978. The evolution of fluvial style, with special reference to the central Appalachian Paleozoic. In *Fluvial sedimentology,* ed. A. D. Miall, 361–84. Calgary: Canadian Society of Petroleum Geologists.

Croft, W. N. 1952. A new *Trochiliscus* (Charophyta) from the Downtonian of Podolia. *Brit. Mus. (Nat. His.) Geol. Bull.* 1:189–220.

Cummins, K. W. 1975. The importance of different energy sources in freshwater ecosystems. In *Productivity of world ecosystems,* ed. D. E. Reichle et al., 50–54. Washington: National Academy of Sciences.

Denison, R. H. 1956. A review of the habitat of the earliest vertebrates. *Fieldiana, Geol.* 11(6):367–457.

———. 1961. Feeding mechanisms of Agnatha, Acanthodii and Placoderma. *Am. Zool.* 1:171–81.

———. 1979. *Acanthodii.* Stuttgart: Fischer.

Donovan, R. N. 1980. Lacustrine cycles, fish ecology and stratigraphic zonation in the middle Devonian of Caithness. *Scott. J. Geol.* 16:35–50.

Douglas, J. G., and A. Lejal-Nicol. 1981. Sur les premières flores vasculaires terrestres du Silurien: Une comparaison entre la "Flore á *Baragwanathia:* d'Australia et la Flore à Psilophytes" et "Lycophytes: d'Afrique du Nord." *Acad. Sci., Paris, C. R., Ser. II* 292:685–88.

Drewry, G. E., A. T. S. Rumsey, and A. G. Smith. 1974. Climatically con-trolled sediments, the geomagnetic field, and trade wind belts in Phanerozoic time. *Geol.* 82:531–53.

Edwards, D. 1980. Early land floras. In *The terrestrial environment and the origin of land vertebrates,* ed. A. L. Panchen, 55–85. New York: Academic Press.

Edwards, D., H. G. Bassett, and E. G. W. Rogerson. 1979. The earliest vascular land plants: Continuing the search for proof. *Lethaia* 12:33–124.

Edwards, D., and E. C. W. Davies. 1976. Oldest recorded *in situ* tracheids. *Nature* 263:494–95.

Edwards, D., and J. Feehan. 1980. Records of Cooksonia-type sporangia from late Wenlock strata in Ireland. *Nature* 287:41–42.

Fischer, W. A. 1978. The habitat of the early vertebrates: Trace and body fossil evidence from the Harding Formation (Middle Ordovician), Colorado. *Mt. Geol.* 15:1–26.

Fisher, T. R., P. R. Carlson, and R. T. Barber. 1982. Sediment nutrient regeneration in three North Carolina estuaries. *Estuarine, Coastal, Shelf Sci.* 14:101–16.

Garrels, R. M., A. Lerman, and F. T. Mackenzie. 1976. Controls of atmospheric O_2 and CO_2: Past, present, and future. *Am. Sci.* 64:306–15.

Garrett, M. J. 1979. New evidence for a Silurian (Ludlow) age for the earliest *Baragwanathia* flora. *Alcheringa* 2:217–24.

Golubic, S., and S. E. Campbell. 1979. Analogous microbial forms in recent subaerial habitats and Precambrian cherts. *Precambrian Res.* 8:201–17.

Gray, J., and A. J. Boucot. 1977. Early vascular land plants: Proof and conjecture. *Lethaia* 10:145–74.

———. 1978. The advent of land plant life. *Geology* 6:489–92.

Gray, J., D. Massa, and A. J. Boucot. 1982. Caradocian land plant microfossils from Libya. *Geology* 10:197–201.

Grime, J. P. 1977. Evidence for existence of three primary strategies in plants and its relevance to ecological and evolutionary theory. *Am. Nat.* 111:1169–94.

———. 1979. *Plant strategies and vegetation processes*. New York: Wiley.

Halstead, L. B. 1973. The heterostracan fishes. *Biol. Rev. Cambridge Philos. Soc.* 48:279–332.

Heckel, P. H., and B. J. Witzke. 1979. Devonian world palaeogeography determined from distribution of carbonates and related lithic palaeoclimatic indicators. In *The Devonian system*, ed. M. House, 99–123. London: Palaeontological Association.

Hoffman, P. 1976. Environmental diversity of Middle Precambrian stromatolites. In *Stromatolites*, ed. M. R. Walter, 599–611. New York: Elsevier.

Hofmann, H. J., D. A. B. Pearson, and B. H. Wilson. 1980. Stromatolites and fenestral fabric in Early Proterozoic Huronian Supergroup, Ontario. *Can. J. Earth Sci.* 17:1351–57.

Hopkinson, C. S., and Day, J. W. 1979. Aquatic productivity and water quality at the upland-estuary interface in Barataria Basin, Louisiana. In *Ecological processes in coastal and marine systems*, ed. R. J. Livingston, 291–314. New York: Plenum.

Hoskins, D. M. 1961. Stratigraphy and paleontology of the Bloomsburg Formation. *Bull. — Pa. Topogr. Geol. Sur.*, G36.

Hughes, N. F., and J. Smart. 1957. Plant-insect relationships in Palaeozoic and later time. In *The fossil record*, ed. W. B. Harland et al., 107–18. London: Geological Society.

Jackson. T. A., P. Fritz, and R. Drimmie. 1978. Stable carbon isotope ratios and chemical properties of kerogen and extractable organic matter in pre-Phanerozoic and Phanerozoic sediments — Their interrelations and possible paleobiological significance. *Chem. Geol.* 21:335–50.

Johnson, J. H., and K. Konishi. 1959. Studies of Devonian algae. *Q. Colo. Sch. Mines* 53:1–114.

Kevan, P. G., W. G. Chaloner, and D. B. O. Savile. 1975. Interrelationships of early terrestrial arthropods and plants. *Palaeontology* 18:391–417.

Kidston, R., and W. H. Lang. 1921. On Old Red Sandstone plants showing structure, from the Rhynie Chert Bed, Aberdeenshire. Part V. The Thallophyta occurring in the peat-bed; the succession of plants throughout a vertical section of the bed, and the conditions of accumulation and preservation. *Trans. — R. Soc. Edinburgh* 52:855–902.

Kjellesvig-Waering, E. N. 1961. The Silurian Eurypterida of the Welsh borderland. *J. Paleontol.* 35:789–835.

Lamont, A. 1948. Gala-Tarannon Beds in the Pentland Hills, near Edinburgh. *Geol. Mag.* 84:193–208, 289–303.

Lang, W. H. 1937. On the plant remains from the Downtonian of England and Wales. *Philos. Trans. R. Soc. London, Ser. V*, 227B:245–91.

Laurie, M. 1899. On a Silurian scorpion and some additional euripterid remains from the Pentland Hills. *Trans. — R. Soc. Edinburgh* 39:575–90.

Livingston, R. J., and J. L. Duncan. 1979. Climatological control of a north Florida coastal system and impact due to upland forestry management. In *Ecological processes in coastal and marine systems*, ed. R. J. Livingston, 339–82. New York: Plenum Press.

McKerrow, W. S. 1979. Ordovician and Silurian changes in sea level. *J. Geol. Soc., London* 136:137–45.

Massa, D. 1980. A stratigraphic contribution to the Paleozoic of the southern basins of Libya. In *The geology of Libya*, M. J. Salem and M. T. Busrewil, 1:6–32. London: Academic Press.

Moss, B. 1980. *Ecology of fresh waters*. New York: Wiley.

Nalivkin, D. V. 1973. *Geology of the USSR*. Edinburgh: Oliver and Boyd.

Niklas, K. J. 1976a. Chemataxonomy of *Prototaxites* and evidence for possible terrestrial adaptations. *Rev. Palaeobot. Palynol.* 22:1–17.

———. 1976b. Morphologic and ontogenetic reconstruction of *Parka decipiens* Fleming and *Pachytheca* Hooker from the Lower Red Sandstone, Scotland. *Trans. — R. Soc. Edinburgh* 69:483–99.

———. 1980. Paleobiochemical techniques and their applications to paleobotany. *Prog. Phytochem.* 6:143–81.

Niklas, K. J., and L. M. Pratt. 1980. Evidence for lignin-like constituents in early Silurian (Llandoverian) plant fossils. *Science* 209:396–97.

Niklas, K. J., and V. Smocovitis. 1983. Evidence for a conducting strand in Early Silurian plants: Implications for the evolution of land plants. *Paleobiology* 9:126–37.

Niklas, K. J., B. H. Tiffney, and A. H. Knoll. 1980. Apparent changes in the diversity of fossil plants. *Evol. Biol.* 12:1–90.

Olson, E. C. 1966. Community evolution and the origin of mammals. *Ecology* 47:291–302.

———. 1971. *Vertebrate paleozoology.* New York: Wiley-Interscience.

Pimm, S. L., and J. H. Lawton. 1977. Number of trophic levels in ecological communities. *Nature* 268:329–31.

Pirozynski, K. A., and D. W. Malloch. 1975. The origin of land plants: A matter of mycotrophism. *BioSystems* 6:153–64.

Pratt, L. M., T. L. Phillips, and J. M. Dennison. 1978. Evidence of non-vascular land plants from the early Silurian (Llandoverian) of Virginia, U.S.A. *Rev. Palaeobot. Palynol.* 25:121–50.

Ratner, M. I., and J. C. G. Walker. 1972. Atmospheric ozone and the history of life. *J. Atmos. Sci.* 29:803–08.

Raven, J. A. 1977. The evolution of vascular land plants in relation to supracellular transport processes. *Adv. Bot. Res.* 5:153–232.

Retallack, G. 1981. Fossil soils: Indicators of ancient terrestrial environments. In *Paleobotany, paleoecology and evolution*, ed. K. J. Niklas, 1:55–102. New York: Praeger.

———. N.d. Fossil soils as grounds for interpreting the advent of large plants and animals on land. *Philos. Trans. R. Soc. London*, in press.

Rhoads, D. C., and J. W. Morse. 1970. Evolutionary and ecologic significance of oxygen-deficient marine basins. *Lethaia* 4:413–28.

Roberts, T. R. 1972. Ecology of fishes in the Amazon and Congo basins. *Bull. Mus. Comp. Zool.* 143:117–47.

Rolfe, Ian. 1980. Early invertebrate terrestrial faunas. In *The terrestrial environment and the origin of land vertebrates*, ed. A. L. Panchen, 117–57. New York: Academic Press.

Romer, A. S. 1955. Fish origins — Fresh or salt water? *Pap. Mar. Biol. Oceanogr., Deep Sea Res.*, suppl. to vol. 3, 261–80.

Schopf, J. M., E. Mencher, A. J. Boucot, and H. N. Andrews. 1966. Erect plants in the early Silurian of Maine. *U.S. Geol. Surv. Prof. Paper* 550D:69–75.

Schumm, S. A. 1968. Speculations concerning paleohydrologic controls of terrestrial sedimentation. *Geol. Soc. Am. Bull.* 79:1573–88.

Scotese, C. R., R. K. Bambach, C. Barton, R. Van der Voo, and A. M. Ziegler. 1979. Paleozoic base maps. *J. Geol.* 87:217–77.

Scourfield, D. J. 1940. The oldest known fossil insect. *Nature* 145:799–801.

Sculthorpe, C. D. 1967. *The biology of aquatic vascular plants.* New York: St. Martin's Press.

Shear, W. A., P. M. Bonamo, J. D. Grierson, W. D. Ian Rolfe, E. L. Smith, and R. A. Norton. 1984. Early land animals in North America: Evidence from Devonian Age arthropods from Gilboa, New York. *Science* 224:492–94.

Smart, J., and N. F. Hughes. 1972. The insect and the plant: Progressive palaeoecologic integration. In *Plant-insect relationships*, ed. H. F. van Emden, 143–55. London: Royal Entomological Society.

Smith, D. G. 1979. The distribution of trilete spores in Irish Silurian rocks. In

The Caledonides of the British Isles, ed. A. L. Harris, C. H. Holland, and B. E. Leake. *Geol. Soc. London Spec. Pub.* 8:423–31.

Spjeldnaes, N. 1979. The palaeoecology of the Ordovician Harding Sandstone. *Palaeogeogr., Palaeoclimat., Palaeoecol.* 26:317–47.

Stebbins, G. L., and G. J. C. Hill. 1980. Did multicellular plants invade the land? *Am. Nat.* 115:342–53.

Størmer, L. 1955. Merostomata. In *Treatise on invertebrate paleontology, Part P*, ed. R. C. Moore, 2:4–41. Boulder, CO: Geological Society of America.

———. 1970. Arthropods from the Lower Devonian (Lower Emsian) of Alken an der Mosel, Germany: Part 5. *Senckenbergiana Letheae* 51:335–69.

———. 1976. Arthropods from the Lower Devonian (Lower Emsian) of Alken an der Mosel, Germany: Part 5. *Senckenbergiana Letheae* 57:87–183.

Strother, P. K., and C. Link. 1983. *Eohostimella* is not a plant. *Am. J. Bot.* 70 (no. 5, pt. 2):80.

Swain, T., and G. Cooper-Driver. 1981. Biochemical evolution in early land plants. In *Paleobotany, paleoecology, and evolution*, ed. K. J. Niklas, 103–34. New York: Praeger.

Tasch, P. 1969. Brachiopoda. In *Treatise on invertebrate paleontology, Part R*, ed. R. C. Moore, 4:4–41. Boulder, CO: Geological Society of America.

Tiffney, B. H. 1981. Diversity and major events in the evolution of land plants. In *Paleobotany, paleoecology, and evolution*, ed. K. J. Niklas, 2:193–230. New York: Praeger.

Tiffney, B. H., and K. J. Niklas. In press. The history of clonal growth in land plants; A paleobotanical perspective. In *Population biology and the evolution of clonal organisms*, ed. J. B. C. Jackson and R. Cook. New Haven and London: Yale Univ. Press, forthcoming.

Walker, J. C. G. 1977. Origin of the atmosphere: History of the release of volatiles from the solid earth. In *Chemical evolution of the early Precambrian*, ed. C. Ponnamperuma, 1–11. New York: Academic Press.

Waterston, C. D. 1979. Problems of functional morphology and classification in stylonurid euripterids with observations on the Scottish Silurian Stylonuroidea. *Trans. — R. Soc. Edinburgh* 70:251–322.

Welcomme, R. L. 1979. *Fisheries ecology of floodplain rivers*. London: Longman.

Westoll, T. S. 1950. The vertebrate bearing strata of Scotland. In *International Geological Congress, Report of the 18th Session, Great Britain, 1948*, pt. 11, 5–21.

———. 1979. Devonian fish biostratigraphy. In *The Devonian system*, ed. M. House, 341–53. London: Palaeontological Association.

Whittaker, R. H. 1977. Evolution of species diversity. *Evol. Biol.* 10:1–67.

Ziegler, A. M., R. K. Bambach, J. T. Parrish, S. F. Barrett, E. H. Gierlowski, W. C. Parker, A. Raymond, and J. J. Sepkoski, Jr. 1981. Paleozoic biogeography and climatology. In *Paleobotany, paleoecology and evolution*, ed. K. J. Niklas, 2:231–66. New York: Praeger Press.

Ziegler, A. M., K. S. Hansen, M. E. Johnson, M. A. Kelly, C. R. Scotese, and

R. Van der Voo. 1977. Silurian continental distributions, paleogeography, climatology, and biogeography. *Tectonophysics* 40:473–502.

Ziegler, A. M., C. R. Scotese, W. S. McKerrow, M. E. Johnson, and R. K. Bambach. 1979. Paleozoic paleogeography. *Annu. Rev. Earth Planet. Sci.* 7:473–502.

STEPHEN F. BARRETT

Early Devonian Continental Positions and Climate: A Framework for Paleophytogeography

Introduction

The aim of this chapter is to examine the relationships between paleogeography, paleobiogeography, and paleoclimatology and to show that in reconstructing ancient continental positions, climates, or biogeographies, an integrated approach is the most likely to succeed. A second goal is to show that, despite many uncertainties about the Paleozoic world, it is possible to create useful models of ancient climates and potential biogeographic provinces.

Specific goals are (1) examination and comparison of various proposed paleogeographic reconstructions, each based on different criteria; (2) demonstration of the techniques used in creating a model of climate for any assembly of continents; (3) modeling of Early Devonian climate on four different paleogeographic reconstructions; (4) selection of the continental reconstruction most consistent with different kinds of paleogeographic evidence, including climatically sensitive sediments; and (5) using a paleoclimatic model for the best reconstruction as a basis for the prediction of very generalized terrestrial biogeographic regions for the early Devonian (assuming climate and geography are important determinants of distribution of terrestrial organisms).

The discussion proceeds from topics of an empirical, geologically supported nature to topics that are more derived and theoretical. The first topic is paleogeography, the sources of information for reconstructions, and problems associated with them. The second topic is paleoclimatology and ways of modelling ancient climates. Paleoclimatic models for Early or Middle Devonian reconstructions prepared by four sets of authors illustrate this process. The third topic concerns evidence in support of the paleoclimatic models and continental reconstructions. The last topic deals with predicted paleobiogeography, based on the best reconstruction, and evidence for that paleobiogeography from preliminary multivariate analysis of Early Devonian macrofloras.

This account differs from previous discussions of Devonian paleogeography in that it compares several different reconstructions, including one proposed here for the first time. In addition to providing models of paleoclimate for each reconstruction, this chapter also uses the relationship between climate, geography, and biogeography to predict major Devonian terrestrial biogeographic patterns.

Reconstructions of Ancient Continental Positions

Because of the effects of continental distribution on atmospheric and oceanic circulation and on animal and plant distributions, any sort of global paleobiogeographic or paleoclimatologic investigation requires an accurate base map. The sources of information used for making continental reconstructions are listed below, along with some of the limitations of each technique. In using this information for making reconstructions, multiple lines of evidence are necessary for reliable paleocontinental reconstructions. Anomalous evidence can then be resolved, either as incorrect or as a true anomaly requiring explanation.

Paleomagnetism

The inclination of the earth's magnetic field is a function of latitude and can be preserved by magnetic minerals in sediments during deposition or in igneous rocks during cooling and crystallization (McElhinny, 1973). If a rock sample is restored to its original (depositional) horizontal position and the magnetic effects of postdepositional diagenetic, thermal, and chemical processes are removed, then the paleoinclination of the earth's magnetic field can be measured. For this information to be useful, the precise age of the rock must be known. Further, many measurements are needed for any stratigraphic unit, in order to average out the effect of secular variation (difference in location of the magnetic and geographic poles). Reliable paleopoles generally have a 95 percent level of confidence at $\pm 5-10°$. Figure 5.1 is an example of a paleogeographic reconstruction based on paleomagnetic evidence. Paleomagnetic data can provide information only about

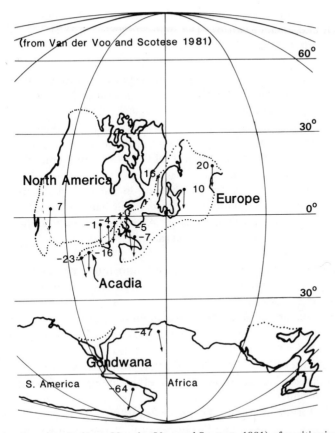

Fig. 5.1. An example (from Van der Voo and Scotese, 1981) of positioning paleocontinents by means of paleomagnetic data. The dots with arrows represent paleomagnetic samples; the arrows point to the Devonian magnetic pole recorded in each sample, and the adjacent numbers indicate the calculated paleolatitudes (negative numbers indicate south latitudes).

the paleolatitude and orientation of continents; no constraints are imposed on longitudinal (east-west) position.

Climatically Sensitive Sediments

The production of certain sediments is strongly controlled by some elements of climate, such as temperature and net evaporation. Heckel and Witzke (1979) and Ziegler, Barrett, and Scotese (1981) discussed the relationships between climatic zones and sedimentation in both modern and ancient worlds. The major features of those relationships are summarized below.

1. Thick carbonates (skeletal and especially nonskeletal in origin) accumulate in the warm tropics and subtropics, within 30–35° of the equator. Coral reefs, because of the light requirements of the symbiotic algae in hermatypic corals, have a parallel distribution. Thick carbonate sequences are largely temperature-dependent, whereas reef corals are both temperature- and insolation-dependent.

2. Thick clastic sediments develop in areas of moderate to high rainfall. Two midlatitude belts occur between roughly 35° and 55° north and south latitude. An equatorial belt of high rainfall and thick clastics is generaly restricted to within 5° of the equator.

3. Most extensive evaporites (dominantly gypsum and halite) accumulate in the low-latitude dry belts, within 15–25° north and south latitude.

4. Coal and peat deposits form in humid zones, in both temperate and equatorial rainy belts. Most coals known today were deposited in the temperate zones (Ziegler, Barrett, and Scotese, 1981; Parrish, Ziegler, and Scotese, 1982); presumably these occurrences are an effect of preservation rather than production.

5. During major glacial intervals, tillites are deposited at high latitudes, although Pleistocene tills occur as far south as 40°N.

Like the climatic zones that influence them, these sediment types do not form bands parallel to latitude, but instead are inclined to lines of latitude.

Inclined zonation is in part the result of climatic modification caused by the oceanic circulation. On west sides of oceans, the warm western boundary current affects adjacent seas and lands and maintains a more humid climate. Conversely, on the east sides of oceans, high ocean-surface temperatures and humid conditions are compressed close to the equator by the cold return portion of major oceanic gyres and by coastal upwelling. This asymmetry is clearly demonstrated by carbonates and coral reefs. In both the Pacific and Atlantic oceans, reefs and abundant carbonates occur as far poleward as 30° north and south latitude on the west sides, but their ranges are restricted to within 5–10° of the equator on the east sides. This distributional asymmetry makes these indicators slightly less useful for determining paleolatitude, but they may aid in inferring the orientation of paleocontinents.

Usefulness increases as a greater variety of indicators are used and with wide geographic coverage over a continent (especially if the continent covered several climatic zones). In general, the most useful paleoclimatic indicators are carbonates and reefs, evaporites, and to a lesser extent coals. The conditions governing their formation are more restrictive than those of other sediments such as thick clastics, and their description and identification are less equivocal than the identification of tillites. Like paleomagnetic data, climatic indicators provide information only about approximate paleolatitude.

Paleobiogeography

Ancient distributions of animals and plants provide paleogeographically useful information. Unlike the preceding two sources of information, biogeography gives more than just paleolatitude; it can give relative longitudinal positions of paleocontinents. The basis for using paleobiogeography in this way is that large contiguous areas often contain a distinct fauna or flora. More than one continent may share such a biogeographic realm or region; the sharing of a paleobiogeographic region by different areas can indicate original contiguity or proximity of those areas.

The major determinants of such biotic distributions are mobility (or barriers to mobility), and similarity and suitability of environments. For most of the biota, especially sedentary forms such as plants, corals, bivalves, or brachiopods, the greatest mobility is attained in larval (or seed) dispersal (Scheltema, 1977). In general, barriers to dispersal result from some combination of distance and intervening hostile environments. The effect of a barrier may vary from complete to partial, and the effect may vary from taxon to taxon.

Barriers to dispersal of marine forms are land (such as the Central American isthmus), deep ocean basins (for continental-shelf dwellers), and steep temperature gradients. For terrestrial organisms, barriers include wide oceans, high mountain ranges, and deserts.

Biogeographic information is useful to paleogeography in two ways. First, high endemism — the presence of a large proportion of taxa unique to an area — indicates barriers to dispersal beyond the geographic limits of the endemism. Second, great similarity of faunas or floras in different areas implies dispersal between those areas. In investigating biotic similarities, quantitative methods and multivariate statistics are useful tools; for some examples of the quantitative approach to paleobiogeography see Whittington and Hughes (1972), for Ordovician trilobites; Crick (1978), for Ordovician nautiloids; Savage, Perry, and Boucot (1979), for Devonian brachiopods; and Ziegler, Bambach, et al. (1981), for Devonian and Carboniferous plants.

An example (fig. 5.2) illustrates some of the uses of paleobiogeographic information. Two paleocontinents are shown, each with two separate paleobiogeographic units. Case A, at the top, shows unreconstructed or partially reconstructed continental positions. There, the occurrence of similar biotas on both continents suggests greater proximity between the pieces. Case B alters the reconstruction to reflect that proximity. However, the separation between the two paleobiogeographic units indicates a barrier to dispersal between them. Case C shows the barrier. In ways such as this, paleobiogeography can place major constraints on continental reconstructions.

However, the limitations on paleobiogeographic data are many. The problems inherent in any paleontologic or paleoecologic endeavor

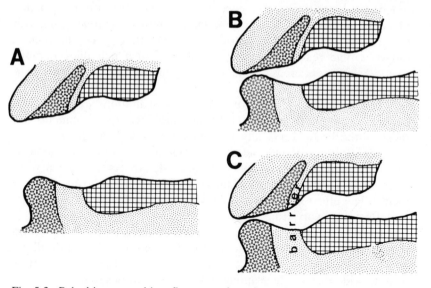

Fig. 5.2. Paleobiogeographic refinement of continental reconstructions. Light stipple indicates land; heavy stipple and grid pattern each represent a distinct marine paleozoogeographic unit.
A. Two paleocontinents prior to reconstruction.
B. Paleocontinents reconstructed closer together, based on faunal similarities.
C. Separation of the two paleobiogeographic units suggests a barrier, as indicated.

are present: adequacy of sampling, stratigraphic correlation, and dating of the deposits. In addition, variability in sampling of different areas must be a concern, as must standardization of taxonomic work carried out over decades and in different countries and schools of taxonomic thought. In comparing the biotas of different areas, comparability of environments represented in samples is also very important (Taylor and Forester, 1979). In addition, the paleogeographic meaning of a barrier depends on the nature of barriers with regard to dispersal for the taxa involved; this may not always be clear. Finally, biotic similarity is not a direct function of geographic proximity. During times of high worldwide cosmopolitanism, similarity is obviously not related to distance. Even during times of high provincialism, similarity is a function of the available means and rates of dispersal, as well as of proximity.

Geologic History

Geologic history is important to paleogeography mostly in the recording of continental collisions and, to a lesser extent, rifting events. Continental collisions are useful because they indicate real proximity, at the time of collision, between two continental pieces. Further, the knowledge that the two pieces were on a collision course permits reconstruction of earlier relative positions (or at least the probable geographic order or sequence) of those pieces. The usefulness of this method is limited as earlier and earlier times prior to the collision are examined.

Devonian Reconstructions and Sources of Information

Before examining Devonian paleoclimatology and paleobiogeography, it is appropriate to consider the nature of the evidence used to make Devonian reconstructions and to look at some of the reconstructions that have been proposed. The four reconstructions discussed are not all for the Early Devonian, since some authors do not include that time interval in their sets of reconstructions. However, reconstructions for Middle Devonian or Late Silurian times can serve as a guide for what their authors' Early Devonian reconstruction would have looked like. For this general discussion, making that assumption is acceptable because the differences in reconstructions resulting from different paleogeographic approaches are much greater than differences in reconstructions resulting from relative movement of continents over a few tens of millions of years.

Paleomagnetic information for the Devonian is summarized in McElhinny (1973) and Scotese et al. (1979). North America has one paleomagnetic determination for the Early Devonian and none for the Middle Devonian; northern Europe and the Russian platform have a total of eleven paleomagnetic determinations for the Early Devonian and one for the Middle Devonian. Siberia has four and seven paleomagnetic determinations for the Early and Middle Devonian, respectively. Gondwana has four Early Devonian paleomagnetic determinations, two in South America and two in Australia (though the Australian poles have been questioned: Scotese et al., 1979). No paleomagnetic data from China are known. The reconstruction in figure 5.3, from Scotese et al. (1979), is based on this paleomagnetic evidence. Van der Voo and Scotese (1981) also discussed the implications of Middle and Late Devonian paleomagnetism with respect to North America, Europe, and Gondwana. They incorporated additional data; figure 5.1 shows their conclusions.

The positions of Gondwana and North America are poorly constrained by paleomagnetic data. The Early and Middle Devonian paleolatitudes of North America can be interpolated with some confidence from Late Silurian and Late Devonian poles; this is not true for Gondwana (Scotese et al., 1979; Hailwood, 1974).

Another reconstruction based on paleomagnetism is that of Smith, Hurley, and Briden (1981) for the Late Devonian (Frasnian), shown in figure 5.4. This is their only Devonian map, but since their Late Silurian map is approximately the same, and assuming that relatively little movement of continents occurred in the interval between the two maps, the Late Devonian reconstruction may be used as an approximation for the Early and Middle Devonian. Note the large circle of error (alpha–95) on the pole for Gondwana. Note also that, unlike Scotese et al. (1979), Smith, Hurley, and Briden (1981) chose to keep all of Asia together. Geologic and paleobiogeographic evidence suggests that throughout most of the Paleozoic, Asia was separated into at least four pieces: Kazakhstania, Siberia, North China, and South China. For further discussion of the geologic evidence, see Ziegler, Hansen, et al. (1977); for the biogeographic evidence, see Ziegler, Bambach, et al. (1981).

Heckel and Witzke (1979) have proposed a Devonian reconstruction based on lithofacies and climatic indicators (fig. 5.5). Although their reconstruction is basically for the Middle Devonian, they plot paleoclimatic evidence from all of the Devonian, with consequent loss of some latitudinal resolution. Their reconstruction differs from the others by placing most continental pieces close together and generally at lower latitudes.

Other reconstructions for the Devonian have been proposed (Kep-

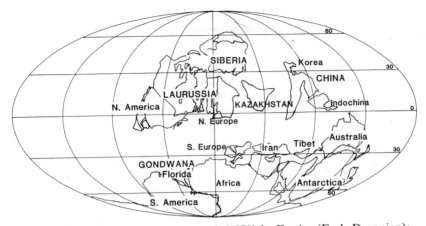

Fig. 5.3. Reconstruction by Scotese et al. (1979) for Emsian (Early Devonian); Mollwiede projection. Paleocontinents are identified by names in capitals, and modern continents and regions by names in lower case. Laurussia is composed of North America and northern Europe.

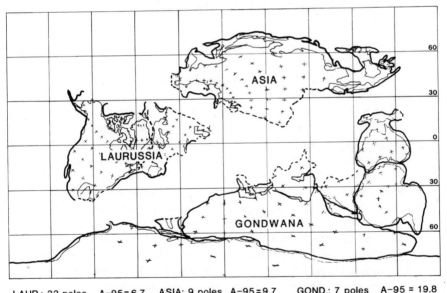

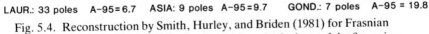

LAUR.: 33 poles   A−95=6.7   ASIA: 9 poles A−95=9.7   GOND.: 7 poles   A−95 = 19.8

Fig. 5.4. Reconstruction by Smith, Hurley, and Briden (1981) for Frasnian (Late Devonian); Mercator projection. At the base of the figure is a summary of paleomagnetic data used to make this reconstruction. For each paleocontinent, the number of poles and alpha−95 (95 percent confidence interval, in degrees, for the mean paleopole for the continent) are listed.

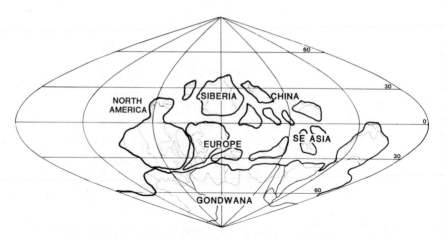

Fig. 5.5. Reconstruction by Heckel and Witzke (1979) for the Middle
Devonian; sinusoidal projection. Present day coastlines and political
boundaries are dotted. For identification of unnamed continental
fragments, see Heckel and Witzke (1979).

pie, 1977; Briden, Drewry, and Smith, 1974; Smith, Briden, and
Drewry, 1973) but are not illustrated here. All are based on
paleomagnetic data, and each is somewhat similar to at least one of the
reconstructions in figures 5.3–5.5.

The major similarity in the three reconstructions presented here is
the latitude of Gondwana; North Africa is placed at roughly 40°S. The
position of North America varies considerably, from 10°N–40°S (Heckel and Witzke, 1979) to 40°N–10°S (Scotese et al., 1979). The position
of Siberia proposed by Heckel and Witzke is also about 30° south of its
position on the two other reconstructions. Because both paleomagnetism and climatically sensitive sediments can provide information only
about paleolatitude, but not about east-west relationships of continents,
the reconstructions by Smith, Hurley, and Briden (1981) and Heckel
and Witzke (1979) have constraints only on paleolatitude. Scotese et al.
(1979) have used paleobiogeographic information (from marine invertebrates) to refine east-west spacing of the paleocontinents.

A fourth reconstruction is proposed here for the Early Devonian
Emsian Age (fig. 5.6). This is a modification of the Scotese et al. (1979)
reconstruction; the position of Gondwana is changed and some rear-
rangements made with small fragments between North America and
Europe.

In this revision, Gondwana is moved northward by 10° on the basis

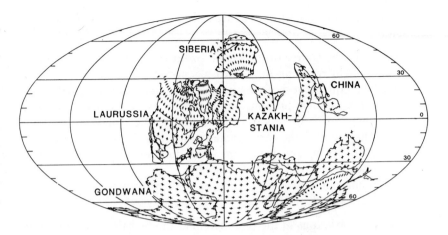

Fig. 5.6. Reconstruction for the Emsian (Early Devonian), proposed here. This is a modification of the Scotese et al. (1979) reconstruction, with Gondwana moved northward and the Acadian terranes moved southward relative to North America. See text for details. The computer-drawn reconstruction was provided by C. R. Scotese. Latitude-longitude ticks have a 5° spacing.

of biogeographic evidence (discussed below), which provides more resolution than the paleomagnetic data. Climatic indicators in Gondwana are consistent with this change. The other rearrangement involves fragments — allochthonous terranes or microplates — which include Acadia (coastal New England and the Canadian Maritimes) and possibly other parts of eastern North America (Van der Voo and Scotese, 1981). These pieces are reconstructed adjacent to North America but far south of their Permo-Triassic position relative to North America. Their northward movement is postulated to have occurred in mid-Carboniferous time, as a result of the collision of Gondwana with North America and Europe (Lefort and Van der Voo, 1981).

Climatology and Global Paleoclimatic Models

Climate (long-term regimes of precipitation and temperature) is the result of a number of interacting factors, which include insolation (a function of latitude), the equator-to-pole temperature gradient, the rotation of the earth, the inclination of the earth's axis of rotation to the plane of its orbit around the sun, and the distribution of land and sea. To simplify the process of modelling paleoclimates, we can make some assumptions about long-term changes in important factors controlling climate.

1. Over the course of the Phanerozoic, insolation (the amount of sunlight reaching the earth) has not changed significantly.
2. The equator-to-pole temperature gradient may have been less drastic than at present, but a gradient existed as a result of the poles receiving less solar heating than the equator.
3. The speed of the earth's rotation has not changed appreciably. The slowing by 10 percent since the Devonian (Wells, 1963) does not constitute a drastic change.
4. Some amount of inclination of the earth's axis to the ecliptic has existed; this affects only seasonality.

Normal variations in the parameters in the first three assumptions will change the intensity of the circulations discussed below, but such variations will not alter the basic pattern (Ziegler, Bambach, et al., 1981). Changes in the fourth parameter affect only the seasonality of the basic pattern. The predictable modifications of this basic pattern by any distribution of land and sea enable the paleoclimatologist to construct models of ancient climate. The following discussion is drawn mainly from Petterssen (1969) and Ziegler, Scotese, et al. (1979).

Atmospheric Circulation

Atmospheric circulation is driven by differential heating between equator and poles. To maintain the global heat balance by convection, cold polar air flows toward the equator and equatorial air flows toward the poles. On a nonrotating earth, the circulation in cross section would consist of surface flow equatorward from the poles and at altitude a reverse flow of warm equatorial air poleward.

However, rotation of the earth introduces additional complexity. Because of the Coriolis effect, air moving northward or southward is deflected to the right in the northern hemisphere and to the left in the southern hemisphere. This results in zonal flow (parallel to lines of latitude) and a consequent system of latitudinal cells, yielding a series of zonal surface highs and lows (fig. 5.7): an equatorial low (warm air ascending), a polar high (dense, cold air descending), and to complete the circuit, a subpolar low-pressure zone and a subtropical high.

Examination of the present-day atmospheric circulation and pressure (fig. 5.8) shows that this idealized general pattern is very distorted. The reason for this distortion is the lack of uniformity of the earth's surface, which is especially important in terms of the thermal contrast between land and sea. Because of the high specific heat of water, large

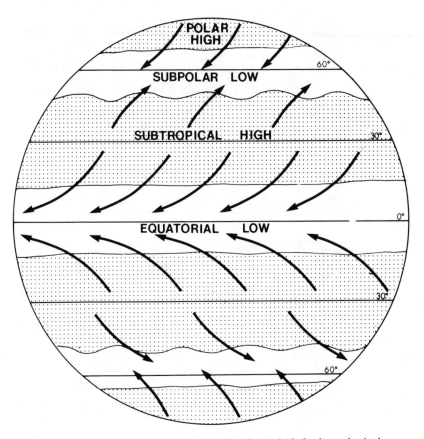

Fig. 5.7. Atmospheric surface pressure and surface winds for hypothetical circulation model on a rotating globe.

bodies of water heat and cool much more slowly than land areas. During the summer, continents become much warmer than the adjacent seas; the rising warm air causes a surface low-pressure area over the land. In the winter the situation reverses, the land becoming relatively cold, which results in high pressure. These reversals are responsible, for example, for the monsoonal circulation (seasonal reversal of flow) between Asia and the Indian Ocean, as shown in figure 5.8.

Oceanic Circulation

Oceanic circulation, though largely driven by atmospheric circulation (Schopf, 1980), is important because it influences climate. For example, midlatitude west coasts of continents are generally dry because the

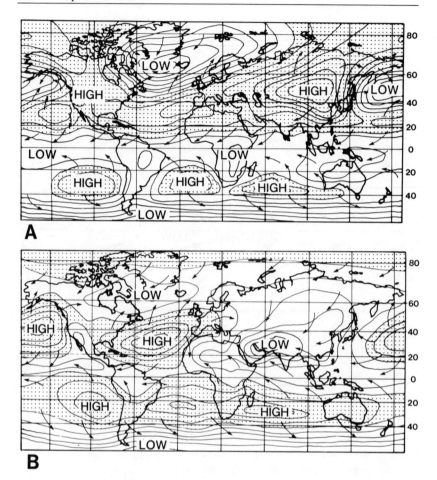

Fig. 5.8. Modern atmospheric circulation (surface winds) and pressure. Areas of relatively high pressure are stippled; arrows indicate wind direction.
A. Northern hemisphere, winter.
B. Northern hemisphere, summer.

return portion of the major oceanic gyres brings cold waters from high latitudes along that portion of the coast. Examples of this current are the Peru, California, and Benguela currents. Onshore winds, which would normally bring moisture-laden air onto the land, are cooled as they pass over the cold current and lose most of their moisture. This accentuates the normal dryness at these latitudes caused by the dry descending air of the low-latitude cell and thus results in the extreme aridity of areas such as the Atacama and Kalahari deserts. Warm western-boundary currents such as the Gulf Stream and the Kuroshio Current also modify the climate of adjacent areas, extending warmer conditions to anomalously high latitudes. Mountainous areas and high plateaus also influence local climate, especially in creating rain shadows. From the point of view of paleogeography and paleoclimatology, only major highlands of a magnitude equivalent to the Himalayas or Tibetan Plateau are important, since their effects may be subcontinental in scope.

Using the principles outlined above — zonal circulation modified by geography, orography, and oceanic circulation — and given a base map showing the distribution of continental masses, it is possible to make a qualitative model of major large-scale pressure distributions and the resultant atmospheric circulation. This model can be used to modify the idealized climatic model (fig. 5.9), which is a consequence of the idealized circulation. For additional examples of this technique, see Robinson (1973). The following section discusses the techniques for

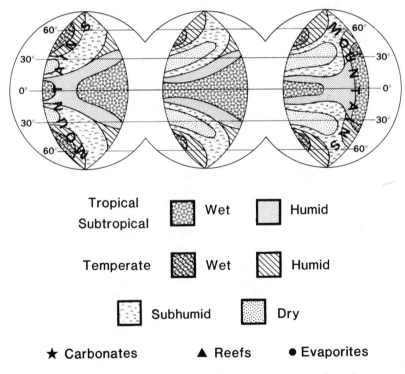

Tropical Subtropical	Wet	Humid
Temperate	Wet	Humid
	Subhumid	Dry

★ Carbonates   ▲ Reefs   ● Evaporites

Fig. 5.9. Generalized climates for an idealized continent (center), and modifications by mountains, as indicated (left and right).

producing models of paleoclimate and relating those models to'a specific set of reconstructions.

Devonian Climatic Models

Using the theoretical considerations discussed above and continental reconstructions in figures 5.3–5.6, we can now produce models of climate. The steps involved in this process are the modeling of (1) winter and summer atmospheric circulation, (2) oceanic circulation, and (3) climatic zones. Because the climate-zone map integrates temperature and precipitation-evaporation, a qualitative precipitation map may be an intermediate step. A set of maps for all these steps will be shown only for the first reconstruction; subsequent reconstructions will be illustrated by only the end product, the climatic model.

Rather than selecting one continental reconstruction as a base for a climatic model, I chose to use each of the four reconstructions in figures 5.3–5.6. In addition to being objective, this approach stresses the need for some skepticism in dealing with any continental reconstruction. Much more important, each of these paleoclimatic and paleogeographic reconstructions can be tested by means of climatic indicators and paleobiogeography.

The climatic zones shown on the figures are tropical-subtropical wet and humid, temperate wet and humid, plus subhumid and dry in any temperature zone. The rainfall categories correspond roughly to modern annual precipitation ranges as follows: wet, more than 150 cm; humid, 100–150 cm; subhumid, 50–100 cm; dry, less than 50 cm. The climatic zones are generalized; their boundaries are not precise and may include inaccuracies of up to hundreds of kilometers at any point. Furthermore, these are *predicted* zones, based only on climatic principles and the location of land in the reconstructions. Figure 5.9 shows the expected distribution of these zones on an idealized continent, with climatic effects of variously placed mountains.

Figures 5.10, 5.11, and 5.12 show, respectively, atmospheric circulation, oceanic circulation, and climatic zones for the reconstruction proposed in figure 5.6. Figures 5.13, 5.14, and 5.15 are the models of climate for the reconstructions in figures 5.3, 5.4, and 5.5, respectively.

The major similarity in the four climatic models is Gondwana: a large southern landmass spanning subtropical, temperate, and polar climates — a sort of "mega-Asia." Because of its size, Gondwana would have developed tremendous summer low-pressure and winter high-pressure systems. As a result, large portions of its eastern and northern midlatitude coasts would have experienced pronounced monsoonal effects, with a wet summer and very dry winter. In all the models, the interior of the continent is shown as dry; marginal mountain ranges would exacerbate this effect. The areal extent and depth of marine arms, such as the Amazonas and Paranaíba basins in South America,

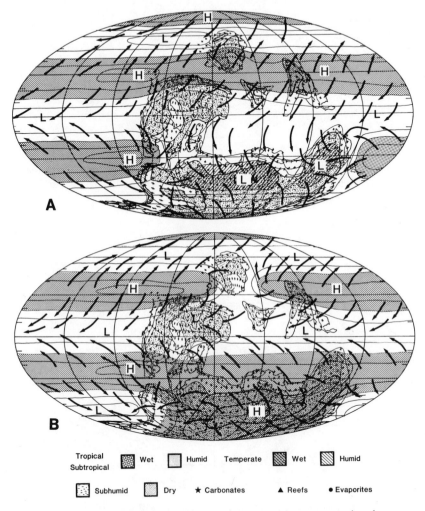

Fig. 5.10. Early Devonian atmospheric circulation and pressure, using the reconstruction in figure 5.6. Land areas are shown in coarse stipple; areas of relatively high pressure are stippled.
   A. Northern hemisphere, winter; the area of extreme low pressure over Gondwana is diagonally lined.
   B. Northern hemisphere, summer.

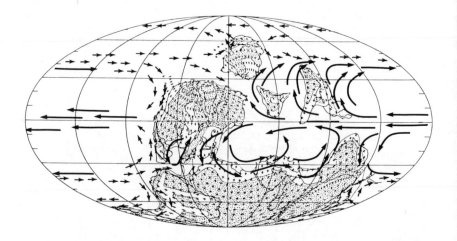

Fig. 5.11. Early Devonian oceanic circulation. Land areas are shown in coarse stipple; long arrows indicate relatively warm-water currents, and short arrows indicate cold-water currents.

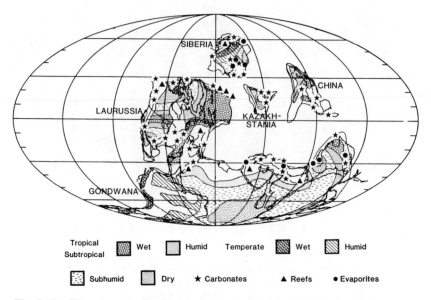

Fig. 5.12. Climatic zones for the reconstruction in figure 5.6; shown only for land areas. See fig. 5.9 for key.

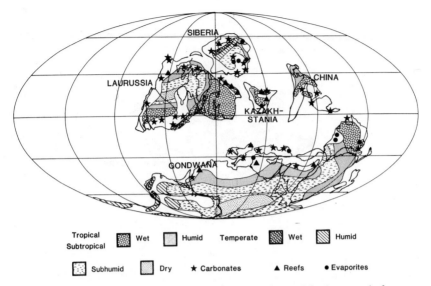

Fig. 5.13. Climatic zones for the reconstruction in figure 5.3, shown only for land areas. See figure 5.9 for key.

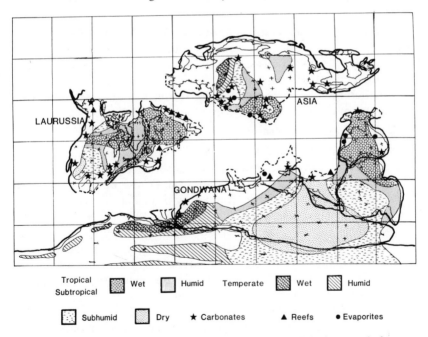

Fig. 5.14. Climatic zones for the reconstruction in figure 5.4, shown only for land areas. See figure 5.9 for key.

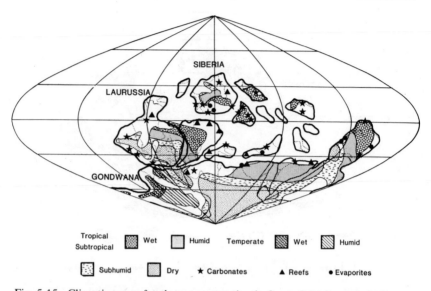

Fig. 5.15. Climatic zones for the reconstruction in figure 5.5, shown only for
land areas. See figure 5.9 for key.

reaching far into the continental interior are other possible factors in
shaping and moderating the climate of the interior of Gondwana. See
Copper (1977) for a discussion of Devonian paleogeography and
climatology in that area.

Differences between the climatic reconstructions, stemming from
the different geographies, are many; the major differences are summarized below.

1. Figures 5.12 and 5.15 suggest a Tethyan-style oceanic circula-
    tion, with a large and partly enclosed equatorial-subtropical
    warm ocean. This circulation would cause warm temperatures
    at anomalously high latitudes, with similar effects on adjacent
    land (Humphreville and Bambach, 1979). Figures 5.13 and
    5.14 maintain the separation between Gondwana and equato-
    rial continents, which would probably cause a reduction in the
    Tethyan effect. Note that figure 5.15 approximates a Pangaean
    reconstruction, and thus the climates shown would be approx-
    imately those for a truly Pangaean Devonian world (see Boucot
    and Gray, 1979).
2. Figure 5.15 shows Siberia in a low-latitude position, causing a
    distribution of theoretically predicted climates on that paleo-
    continent different from those in the other three models.

3. An integral Asia in figure 5.14 implies that China and Indochina would have been cooler and dryer than they are inferred to have been in the other reconstructions.

4. The slightly different positions and orientations of Gondwana result in different climatic zones. Figures 5.13 and 5.14 place Gondwana in higher latitudes and in such a way that the South American–African margin of the continent angles poleward toward the west; figures 5.12 and 5.15 place Gondwana in lower latitudes, with the northern margin parallel to lines of latitude. The main difference in the climatic reconstructions is that in the former case wetter conditions would have prevailed along the northern margin. However, dryer conditions are somewhat masked in the latter case by differences in the geography (more Pangaean) and by the Tethyan circulation mentioned above.

Although the positions of paleocontinents vary from reconstruction to reconstruction, some generalizations are possible concerning continental paleoclimates. The following summary refers more to figures 5.12, 5.13, and 5.15 than to figure 5.14, which keeps Asia together (see the discussion above).

In general, Laurussia would have had a tropical to subtropical climate, with possibly dry areas in the northwest and southwest (northern Canada and southwestern United States). The eastern part (Europe) was probably wet, although the presence of the Acadian mountains and any high terrain on the eastern margin may have created a nonuniform pattern of more locally controlled climates, including subhumid to dry areas.

Kazakhstania and China would have had subtropical to warm-temperate climates. Dry areas could have developed in the northwestern parts of both pieces, depending largely on their paleolatitudes.

The climate of Gondwana would have been dry and cold (polar to cold-temperate) in the interior, and warm-temperate to cold-temperate wet or humid along the margins. The northeastern corner (northern and eastern Australia) would have been warmest and wettest; the northern margin (including India and North Africa) would have had a seasonally humid climate, probably warm-temperate. Western Gondwana (South America) was most likely cool-temperate and humid to wet. The eastern extent of moderate to high rainfall would have depended in part on the area of ice-free marine water.

Assuming a high-latitude position of Siberia, as suggested by paleomagnetic data, the expected paleoclimate would have ranged from warm-temperate to cold-temperate, with dry areas developed on the southwest (present-day northeastern Siberia west of the Ver-khoyansk range). As discussed above, the climatic indicators indicate a much warmer climate than is suggested by paleomagnetic evidence.

The paleoclimatic models presented here are generalized models of climate; they are based on continental positions and simple concepts of atmospheric and climatic dynamics. Such models are predictive, test-able, and largely free from circular reasoning, and therefore they can be very useful to paleobotanists, paleontologists, and paleogeographers. With considerably more data and time, more refined climatic models can be produced. The next step in this process (where the theoretical model is the first step) is the inclusion of detailed paleogeography, such as the inferred position of mountains and highlands and other features that modify precipitation patterns. Determining the location and extent of such features requires detailed analysis of stratigraphy and lithofa-cies. The third cycle might integrate known distributions of climatic indicators, keeping in mind that local effects will not be predictable. Further refinements, such as albedo (reflectance of the earth's surface), can be added, but at this level a computer-based modeling approach becomes desirable (see, for example, Barron, Sloan, and Harrison, 1981).

Supporting Evidence for Climatic and Paleogeographic Reconstructions

The main sources of confirming or contradicting evidence for paleogeographic and climatic reconstructions are climatically sensitive sediment types and paleobiogeography. Sediment types (climatic indica-tors), notably carbonates, reefs, and evaporites, are the most definitive. Because a climatic model depends on and stems from a continental reconstruction, the distribution of climatically sensitive sediments that fit a model of climate also support the continental reconstruction on which that model is based. In cases where the distribution of sediments does not fit the model, the magnitude of the difference between prediction and geologic reality indicates whether the climatic model alone is wrong or whether geographic reconstruction, climatic model, or possibly both are wrong. For example, if in a reconstruction a paleocontinent occupies polar latitudes, but carbonates and reefs of that age are widespread on the continent, then the reconstruction is probably wrong. However, if coal swamps inhabited an area predicted to have had a subhumid climate, it is more likely that the climatic model is slightly in error.

The paleobiogeography, which for the Early and Middle Devonian is restricted to marine zoogeography, provides limited paleoclimatic assistance because it depends on oceanic circulation, which possibly overshadows climate as a controlling factor. However, paleobiogeography is very useful for indicating proximity of areas, especially east-west proximity, for which there is no other source of information.

One example of the use of such evidence concerns Siberia in the Early and Middle Devonian. Paleomagnetic evidence from Siberia (summarized in Scotese et al., 1979, and McElhinny, 1973) suggests paleolatitudes of 30–65°N and indicates that Siberia was inverted, its present northern side facing south (see fig. 5.3). Lithofacies evidence is at odds with such high latitudes: abundant carbonates, including reef-building corals, occur in Mongolia at presumed paleolatitudes of 55–62°N (Ziegler, Bambach, et al., 1981); Devonian evaporites occur widely in present-day northern Siberia, at presumed paleolatitudes of 35–45°N. Biogeography, based mostly on brachiopods, indicates affinities with midlatitude areas such as the Uralian region (eastern part of Northern European or Baltic plate) and also the Canadian Arctic islands (Boucot, 1975; Boucot, Johnson, and Talent, 1969). The paleomagnetic data are clearly anomalous. There are two obvious ways to deal with this contradiction. One is to assume the paleomagnetic data are incorrect and to place Siberia at lower latitudes. Alternatively, if the paleomagnetic poles are correct, then climatic or paleoecologic hypotheses must be provided to explain the lithologic and biotic distributions. One example of such an explanation is that a completely land-free and open north pole, with no land closer than about 30° of latitude, might cause a gentle equator-to-pole temperature gradient in the northern hemisphere and thus in effect broaden and move climatic zones northward. Whether this scenario is possible or sufficient is unclear.

Climatic Indicators

Information on the distribution of reefs and carbonate buildups, thick carbonate sequences, evaporites, and coals comes largely from Heckel and Witzke (1979), Ziegler, Scotese, et al. (1979), and unpublished compilations used for the preparation of maps presented by Ziegler, Scotese, et al. (1979). Unfortunately, the data from Heckel and Witzke (1979) are for the entire Devonian and thus have less resolution than data restricted to the time interval under consideration here, the Early Devonian.

The distribution of climatic indicators is shown on the climatic reconstructions themselves (figs. 5.12–5.15). Carbonates and reefs in North Africa (Hollard, 1967) and evaporites and carbonates in Iran and Afghanistan (Ziegler, Scotese, et al., 1979) are consistent with the lower-latitude position and warm climate of Gondwana shown in figures 5.12 and 5.15. Furthermore, the presence of evaporites in the Australian part of Gondwana is consistent with the paleolatitudes shown on figure 5.14; the dry belt as reconstructed in figure 5.13 would have been too far south. A modification to the climatic model, suggested by the evaporites in Iran, is that the dry belt extended from western Australia westward, as indicated by modern climate as well as by Robinson (1973, fig. 4E). One or two mismatches between a climate model and the occurrence of climatic indicators should not be taken as evidence against a particular reconstruction. Only widespread anomalies should be used in this fashion. Local effects such as rain shadow are possible complicating factors.

Extensive carbonates of Emsian age in North America support a low-latitude position within 30° of the equator for that continent. The abundant and widely distributed reefs and evaporites shown by Heckel and Witzke (1979, figs. 4 and 6) support that general position. However, because those indicators represent a summation for the entire Devonian, and because North America very probably moved noticeably northward during that time (Scotese et al., 1979), they should not be used for support of a detailed Emsian climatic model.

Reefs, carbonates, and evaporites in Siberia are at odds with the high latitudes shown on all reconstructions except figure 5.15. This raises serious questions about the paleomagnetic poles used to position Siberia at high latitudes. It is possible to explain away evaporites, but carbonates and reefs are very much at odds with paleomagnetic data.

The low-latitude positions of other major paleocontinents, such as Kazakhstania and China (figs. 5.12, 5.13, and 5.15), are supported by the presence of thick carbonate sequences.

Paleozoogeography

Several recent and detailed discussions of Devonian zoogeography are available: Boucot (1975); House (1979); Norris (1979); and Savage, Perry, and Boucot (1979), although they are largely based on brachiopods. Certain other phyla are useful biogeographically and tend to have patterns of distribution that parallel those of the brachiopods. These taxa include rugose corals (Oliver, 1977, 1976), trilobites (Eldredge and Ormiston, 1979; Ormiston, 1972), and, to a certain extent, echinoderms (Witzke, Frest, and Strimple, 1979).

The three major Early Devonian biogeographic units and their subdivisions are listed below and shown in figure 5.16.

1. Eastern Americas Realm (EAR): Appohimchi (Appalachian Basin), Michigan Basin–Hudson Bay lowlands (Koch, 1981); "Amazon-Colombian" (northwestern South America).
2. Old World Realm (OWR): Nevadan, Cordilleran (western North America), Franklinian (Canadian Arctic islands), Uralian, Rhenish-Bohemian (Europe and North Africa), Tasman (eastern Australia and New Zealand).
3. Malvinokaffric Realm (MR): undivided.

As noted, the major units are labeled *realm*, and the subdivisions correspond to regions or provinces. However, there is a lack of uniformity and objectivity in the definition of paleobiogeographic units, such as in the hierarchy *realm, region, province.* Paleobiogeographers often use the term *province* for a paleobiogeographic entity, but many such provinces are probably the equivalent of modern regions or even realms (Ziegler, Bambach, et al., 1981). Therefore, in the discussion that follows terminology for paleobiogeographic units will be kept deliberately vague.

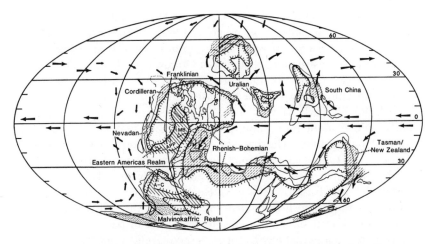

Fig. 5.16. Emsian-Eifelian paleobiogeography and oceanic circulation. *Heavy line with tick marks*: Devonian shoreline; *light line*: edge of Devonian continental shelf; *dashed pattern*: Old World Realm (subunits labeled); *lined pattern*: Eastern Americas Realm (MB = Michigan Basin–Hudson Bay Lowlands subunit, APP = Appohimchi subunit, A-C = Amazon-Colombia subunit); *stippled pattern*: Malvinokaffric Realm; *arrows*: warm currents; *small arrows*: cold currents.

The range of water temperatures usually attributed to the three realms can be inferred from association with reefs, thick carbonates, carbonate-free clastics, or evaporites. The OWR was probably restricted to warm waters, based on the abundance of reefs and carbonates in that realm; the climates involved were tropical to warm-temperate. The EAR faunas inhabited warm waters as well, in tropical to warm-temperate climates. However, this realm may have extended into cool-temperate regions, since EAR faunas in northwestern South America occur in clastic sequences with essentially no carbonates. The MR consisted of cool-temperate to polar marine faunas.

The paleoclimatologic implications of the paleobiogeography are limited, since the temperature limits of the realms are derived from climatic indicators, and there is a danger of circularity of argument. The paleogeographic implications are thus perhaps more important.

The OWR units inhabited warm waters and were within dispersal range of each other, either by continuity of coastline or proximity of the separate pieces. This implied distribution is in agreement with the reconstructions, except for those which show Siberia at high latitudes. The differentiation within the OWR is probably the result of the wide areal extent of the realm and the difficulties in maintaining contact over such large areas.

The EAR faunas should not be expected to extend to high latitudes, since the realm includes large equatorial areas. The presence of EAR faunas in northern South America is therefore interesting, since in any of the four reconstructions that part of Gondwana is likely to have been cool. The distribution of the MR faunas is in accord with the high-latitude position shown in the reconstructions.

Although Early Devonian biogeographic patterns do not provide much paleoclimatic information, the patterns do impose constraints on the paleogeography. The paleobiogeography shown schematically in figure 5.2A is in fact the situation that existed for Early Devonian marine invertebrates in Laurussia and Gondwana. Unit 1 (heavy stipple) is the Eastern Americas Realm, and Unit 2 (crosshatched pattern) is the Rhenish-Bohemian subdivision of the Old World Realm. Since both Laurussia and Gondwana shared both paleobiogeographic units, paleocontinental reconstructions should reflect this faunal similarity as geographic proximity. Because on both paleocontinents the biogeographic units are separated, a barrier must have existed between the units. Between North America and Europe, this barrier was land; the barrier may have extended as a land bridge to Gondwana. Unfortunately, the subsequent late Paleozoic orogenic activity has obscured the geologic record of the southern part of the barrier. The appearance of some Appalachian taxa in North Africa (Drot, 1966; Le Maitre, 1955; Oliver, 1977) suggests that a marine connection existed at least ephemerally between the two paleobiogeographic units.

Thus, the paleobiogeographic pattern provides constraints on the paleogeography of the Early Devonian. This information is in addition to but not completely at odds with paleomagnetic data from Gondwana, which, as noted above, only loosely constrain that paleocontinent's position. It is this paleobiogeographic evidence which is used in making the reconstruction shown in figures 5.6 and 5.12.

Predicted Terrestrial Biogeography

Using a model of ancient climate and geography, it is possible to predict general biogeographic patterns for that model. Of course, this predictive approach requires certain assumptions:

1. Taxa for which the distribution is predicted are sensitive to and controlled by climate.
2. Occurrence of taxa in a region is a function of dispersal and the suitability of environments in that region. Further, a taxon will occur in an area if the climate is appropriate and if the taxon is capable of dispersal to that area.

One predicted paleobiogeography for land areas based on the paleogeography and the climate model in figure 5.12 will be discussed here. However, this is only one of many possible biogeographies. For this example, the criteria established for prediction are very simple: areas in the same climate-zone (tropical-subtropical wet, for example) will have similar biotas if they are within dispersal distance of each other. In the example, this distance is arbitrarily set at 2,000 km. In addition, only wet and humid areas are considered and are here treated together as one climatic zone. This last set of simplifications reflects the greater likelihood of preservation of florules or faunules from wetter areas and the rough nature of the location of climate-zone boundaries.

The general approach taken in using these criteria to predict ancient biotic distributions is that areas separated by more than 2,000 km would have been isolated. As a result, these areas would probably have contained distinct biotas representing separate biogeographic units.

Thus, in figure 5.12, northeastern Gondwana — eastern Australia — has a warm and wet climate that is separated by more than 2,000 km from areas with a similar climate. For that reason, that area is shown in figure 5.17 as a separate biogeographic unit. Conversely, areas such as

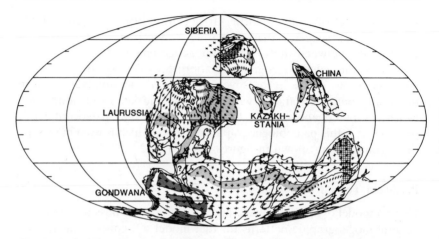

Fig. 5.17. Predicted terrestrial paleobiogeography. See text for discussion. Different patterns indicate areas that would have had distinct biotas.

China, Kazakhstan, central northern Gondwana, and eastern Laurussia have warm wet regions, none of which is more than 2,000 km away from another such region within that group. This fact implies that all these areas should have shared a similar terrestrial biota, as shown in the figure. Varying the criteria slightly — for example, using 1,500 km as the minimum separation required for biogeographic isolation — produces results very similar to those in figure 5.17. In the case of a criterion of 1,500 km separation, China would be predicted to have had a terrestrial biota distinct from European, North African, and North American biotas.

Many factors could be considered in a less general case than that discussed here. These factors include means of dispersal, speed of dispersal, length of viable period of propagules, the degree to which taxa are euryoecious, and probability of successful colonization. The cases presented here are intended as much as demonstrations of an approach as actual predictions of paleobiogeography, although they should have some validity in a general sense. However, the climatic models presented here (or others produced for any other time) can be used to make predictions more tailored to a particular taxon or group of taxa.

Evidence for Devonian Terrestrial Biogeography

The models of paleoclimate and terrestrial paleobiogeography proposed here are theoretical; the only sources of actual geologic information are those used to make the initial continental reconstructions. Furthermore, the proposed terrestrial biogeography is based on simple (if not simplistic) assumptions. Despite these potential problems, preliminary evidence from multivariate analyses of early Devonian phytogeography (Raymond, Parker, and Barrett, this volume) suggests that the modeling process discussed here is a reasonable first approximation of Devonian terrestrial biogeography. The data in these analyses are drawn from summaries by Chaloner and Sheerin (1979) and Banks (1979), with additions from Russian and Chinese publications. For the Early Devonian, there are 46 samples (florules) and 18 genera. As discussed by Raymond, Parker, and Barrett (this volume), the sources of data have been carefully examined so as to include only Early Devonian florules and to standardize the taxonomy as much as possible.

Raymond, Parker, and Barrett (this volume) and Ziegler, Bambach, et al. (1981) have discussed the use of multivariate techniques

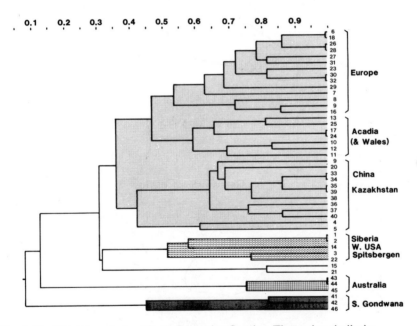

Fig. 5.18. Cluster analysis of Early Devonian florules. The cosine similarity index (equivalent to the Otsuka index, for presence-absence data) and the unweighted pair-group method (UPGM) were used. Level of clustering (similarity) is indicated at the top. The major clusters are shaded; within the large cluster at the top, subclusters are indicated by brackets at the right. See Raymond, Parker, and Barrett (this volume) for locality information and floral lists.

with paleophytogeographic data sets. Interested readers should consult those studies for background on the application of these techniques to paleophytogeography.

The cluster analysis (fig. 5.18) shows two important features. First, the presence of at least six distinct clusters indicates floral differentiation in the Early Devonian. Second, several of the clusters represent florules from areas predicted to have had separate paleobiogeographic units — Australia and southern (including western) Gondwana. In addition, the large tropical-subtropical area, predicted to have had similar biotas by virtue of proximity of the climate-zone regions, does indeed contain groups of florules that are more similar to one another than to the other clusters.

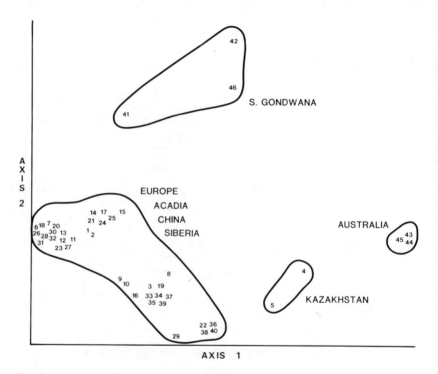

Fig. 5.19. Polar ordination plot of Early Devonian florules. The Dice similarity index was used, with the computer program POLAR II by J. J. Sepkoski, Jr., and J. Sharry. Three axes account for 59 percent of the variability in the similarity matrix; axes 1 and 2 account for 48 percent. The major clusters and subclusters in figure 5.18 are identified by name. See Raymond, Parker, and Barrett (this volume) for locality information and floral lists.

The polar ordination plot (fig. 5.19) shows the same patterns as the cluster analysis; however, here the distinctness of the Australian and southern Gondwanian floras is stressed. The similar tropical-subtropical units form one large, tight group, low on axes 1 and 2, indicating their relatively high similarities, as contrasted to Australia or southern Gondwana.

In the cluster analysis, Siberian florules plus those from Spitsbergen, Greenland, and western North America form a separate and distinct cluster. On the polar ordination plots, these florules are included in the large tropical-subtropical group. This is yet another indication of Siberian affinities with lower-latitude regions. Better sampling of Siberian Devonian floras might reveal the presence of widespread endemics, which could act to differentiate Siberia as a paleobiogeographic unit. However, based on the data available now, Siberia clusters with tropical-subtropical florules.

The results of the multivariate analyses do not conclusively support the climate models. However, these analyses do suggest that the models of paleoclimate may be used to achieve a good first approximation of terrestrial paleobiogeography. By using more refined models of climate (for example, including orogenic and other features) as well as more consideration of the dispersal ability and ecology of the pertinent taxa, the approach outlined here should produce more detailed phytogeographic predictions. Conversely, with a better understanding of Devonian continental positions and more refined paleoclimatic models, multivariate phytogeographic analyses can permit inferences about the dispersal ability of middle Paleozoic plants.

Conclusions

Geography, climate, and biotic distributions are intimately related. As a result, it is possible to model climate for times in the geologic past, given an adequate paleocontinental base map.

If a particular model of global paleoclimate is consistent with independently derived, climatically sensitive geologic evidence (distribution of climatically controlled sediments and paleobiogeography), that model may be used as a tool for predicting distributions of other climatically sensitive items, such as land plants.

A reconstruction of Early Devonian paleoclimate, consistent with paleogeographic information, paleobiogeography, and climatic indicators, suggests that Devonian terrestrial organisms should show differentiation on a global scale. This suggestion is based on the assumption that Devonian land plants and animals had the same general controls on distribution as do modern plants and animals. Barriers to dispersal of such organisms include major sharp temperature gradients and inhospitable wastes, such as deserts, wide oceans, or major mountain ranges.

Preliminary evidence from multivariate analyses of Devonian macrofloras suggests that such biogeographic differentiation, resulting from the combined effects of geography and climate, does exist.

Acknowledgments

A. L. Raymond, A. M. Ziegler, C. R. Scotese, and W. C. Parker have greatly contributed to this chapter with their comments and criticisms. Many of the basic paleogeographic and paleoclimatologic ideas or refinements have been influenced by A. M. Ziegler and the Paleogeographic Atlas project at the University of Chicago. C. R. Scotese kindly provided computer-generated base maps for some of the reconstructions presented here. A. L. Raymond provided paleobotanical insight as well as the multivariate analyses of Early Devonian macrofloras.

References

Banks, H. P. 1979. Floral assemblage zones in the Siluro-Devonian. In *Biostratigraphy of fossil plants: Successional and paleoecological analysis*, ed. D. Dilcher and T. N. Taylor, 1–24. Stroudsburg, PA: Dowden, Hutchinson, Ross.

Barron, E. J., J. L. Sloan, and C. G. A. Harrison. 1981. Potential significance of land-sea distribution and surface albedo variations as a climatic forcing factor; 180 m.y. to the present. *Palaeogeogr., Palaeoclimatol., Palaeoecol.* 30:17–40.

Boucot, A. J. 1975. *Evolution and extinction rate controls*. New York: Elsevier.

Boucot, A. J., and J. Gray. 1979. Epilogue: A Paleozoic Pangaea? In *Historical biogeography, plate tectonics, and the changing environment*, ed. J. Gray and A. J. Boucot, 465–84. Corvallis: Oregon State Univ. Press.

Boucot, A. J., J. G. Johnson, and J. A. Talent. 1969. Early Devonian brachiopod zoogeography. *Geol. Soc. Am., Spec. Pap.* 119.

Briden, J. C., G. E. Drewry, and A. G. Smith. 1974. Phanerozoic equal-area world maps. *J. Geol.* 82:555–74.

Chaloner, W. G., and A. Sheerin. 1979. Devonian macrofloras. In *The Devonian system*, ed. M. R. House, C. T. Scrutton, and M. G. Basset, 145–61. Special Papers in Palaeontology, 23. London: Palaeontological Association.

Copper, P. 1977. Paleolatitudes in the Devonian of Brazil and the Frasnian-Famennian mass extinction. *Palaeogeogr., Palaeoclimatol., Palaeoecol.* 21:165–207.

Crick, R. E. 1978. Ordovician nautiloid biogeography: A probabilistic and multivariate analysis, Ph. D. diss., Univ. of Rochester, Rochester, NY.

Drot, J. 1966. Présence du genre *Amphigenia* (Brachiopode, Centronellidae), dans le bassin de Taoudenni (Maroc). *C. R. Somm. Seances Soc. Geol. Fr.,* 373.

Eldredge, N., and A. R. Ormiston. 1979. Biogeography of Silurian and Devonian trilobites of the Malvinokaffric Realm. In *Historical biogeography, plate tectonics, and the changing environment,* ed. J. Gray and A. J. Boucot, 147–67. Corvallis: Oregon State Univ. Press.

Hailwood, E. A. 1974. Paleomagnetism of the Msissi Norite (Morocco) and the Paleozoic reconstruction of Gondwanaland. *Earth Planet. Sci. Lett.* 23: 376–86.

Heckel, P. H., and B. J. Witzke. 1979. Devonian world palaeogeography determined from distribution of carbonates and related lithic palaeoclimatic indicators. In *The Devonian system,* ed. M. R. House, C. T. Scrutton, and M. G. Basset, 99–123. Special Papers in Palaeontology, 23. London: Palaeontological Association.

Hollard, H. 1967. Le Devonian du Maroc et du Sahara Nord-Occidental. In *International symposium on the Devonian system,* ed. D. H. Oswald, 1:203–44. Calgary, Alberta: Alberta Society of Petroleum Geologists.

House, M. R. 1979. Devonian in the Eastern Hemisphere. In *Treatise on invertebrate paleontology,* ed. R. A. Robison and C. Teichert, Part A (Introduction), A183–A217. Boulder, CO: Geological Society of America and Univ. of Kansas Press.

Humphreville, R., and R. K. Bambach. 1979. Influence of geography, climate, and ocean circulation on the pattern of generic diversity of brachiopods in the Permian. *Geol. Soc. Am. Abs. with Prog.* 11:447.

Keppie, J. D. 1977. Plate tectonic interpretation of Palaeozoic world maps (with emphasis on circum-Atlantic orogens and southern Nova Scotia). *Nova Scotia Dept. Mines Paper* 77–3.

Koch, W. F., III. 1981. Brachiopod community paleoecology, paleobiogeography, and depositional topography of the Devonian Onondaga Limestone and correlative strata in eastern North America. *Lethaia* 14:83–103.

Lefort, J.-P., and R. Van der Voo. 1981. A kinematic model for the collision and complete suturing between Gondwanaland and Laurussia in the Carboniferous. *J. Geol.* 89:537–50.

Le Maitre, D. 1955. Découverte du genre *Amphigenia* dans le Synclinal de Tindouf (Sahara occidental). *Acad. Sci., Paris, C. R.* 240:1659–61.

McElhinny, M. W. 1973. *Paleomagnetism and plate tectonics.* Cambridge: Cambridge Univ. Press.

Norris, A. W. 1979. Devonian in the Western Hemisphere. In *Treatise on invertebrate paleontology,* ed. R. A. Robinson and C. Teichert, Part A (Introduction), A218–53. Boulder, CO: Geological Society of America and Univ. of Kansas Press.

Oliver, W. A., Jr. 1976. Biogeography of Devonian rugose corals. *J. Paleontol.* 50:365–73.

————. 1977. Biogeography of Late Silurian and Devonian rugose corals. *Palaeogeogr., Palaeoclimatol., Palaeoecol.* 22:85–135.

Ormiston, A. R. 1972. Lower and Middle Devonian trilobite zoogeography in northern North America. *24th Int. Geol. Congr. (Montreal),* sec. 7, Palaeontology, 594–604.

Parrish, J. T., A. M. Ziegler, and C. R. Scotese. 1982. Rainfall patterns and the distribution of coals and evaporites in the Mesozoic and Cenozoic. *Palaeogeogr., Palaeoclimatol., Palaeoecol.* 40:67–101.

Petterssen, S. 1969. *Introduction to meteorology,* 3d ed. New York: McGraw-Hill.

Robinson, P. L. 1973. Palaeoclimatology and continental drift. In *Implications of continental drift to the earth sciences,* ed. D. H. Tarling and S. K. Runcorn, 1:451–76. New York: Academic Press.

Savage, N. M., D. G. Perry, and A. J. Boucot. 1979. A quantitative analysis of Lower Devonian brachiopod distribution. In *Historical biogeography, plate tectonics, and the changing environment,* ed. J. Gray and A. J. Boucot, 169–200. Corvallis: Oregon State Univ. Press.

Scheltema, R. S. 1977. Dispersal of marine invertebrate organisms: Paleobiogeographic and biostratigraphic implications. In *Concepts and methods of biostratigraphy,* ed. E. G. Kauffman and J. E. Hazel, 73–107. Stroudsburg, PA: Dowden, Hutchinson, Ross.

Schopf, T. J. M. 1979. The role of biogeographic provinces in regulating marine faunal diversity through geologic time. In *Historical biogeography, plate tectonics, and the changing environment,* ed. J. Gray and A. J. Boucot, 449–53. Corvallis: Oregon State Univ. Press.

————. 1980. *Paleoceanography.* Cambridge: Harvard Univ. Press.

Scotese, C. R., R. K. Bambach, C. Barton, R. Van der Voo, and A. M. Ziegler. 1979. Paleozoic base maps. *J. Geol.* 87:217–77

Smith, A. G., J. C. Briden, and G. E. Drewry. 1973. Phanerozoic world maps. In *Organisms and continents through time,* ed. N. F. Hughes, 1–42. Special Papers in Palaeontology, 12. London: Palaeontological Association.

Smith, A. G., A. M. Hurley, and J. C. Briden. 1981. *Phanerozoic paleocontinental world maps.* Cambridge Earth Science Series. Cambridge: Cambridge Univ. Press.

Taylor, M. E., and R. M. Forester. 1979. Distributional model for marine isopod crustaceans and its bearing on early Paleozoic paleozoogeography and continental drift. *Geol. Soc. Am. Bull.,* Part I 90:405–13.

Van der Voo, R., and C. Scotese. 1981. Paleomagnetic evidence for a large (> 2000 km) sinistral offset along the Great Glen fault during Carboniferous time. *Geology* 9:583–89.

Wells, J. W. 1963. Coral growth and geochronometry. *Nature* 197:948–50.

Whittington, H. B., and C. P. Hughes. 1972. Ordovician geography and faunal provinces deduced from trilobite distribution. *Philos. Trans. R. Soc. London* B263:235–78.

Witzke, B. J., T. J. Frest, and H. L. Strimple. 1979. Biogeography of the Silurian–Lower Devonian echinoderms. In *Historical biogeography, plate tectonics, and the changing environment*, ed. J. Gray and A. J. Boucot, 117–29. Corvallis: Oregon State Univ. Press.

Ziegler, A. M., R. K. Bambach, J. T. Parrish, S. F. Barrett, E. H. Gierlowski, W. C. Parker, A. Raymond, and J. J. Sepkoski, Jr. 1981. Paleozoic biogeography and climatology. In *Paleobotany, paleoecology, and evolution*, ed. K. J. Niklas, 2:231–66. New York: Praeger.

Ziegler, A. M., S. F. Barrett, and C. R. Scotese. 1981. Palaeoclimate, sedimentation and continental accretion. *Philos. Trans. R. Soc. London* A301:253–64.

Ziegler, A. M., K. S. Hansen, M. E. Johnson, M. A. Kelly, C. R. Scotese, and R. Van der Voo. 1977. Silurian continental distributions, palaeogeography, climatology, and biogeography. *Tectonophysics* 40:13–51.

Ziegler, A. M., C. R. Scotese, W. S. McKerrow, M. E. Johnson, and R. K. Bambach. 1979. Paleozoic paleogeography. *Annu. Rev. Earth Planet. Sci.* 7:473–502.

ANNE RAYMOND, WILLIAM C. PARKER, AND
STEPHEN F. BARRETT

Early Devonian Phytogeography

Introduction

In this chapter we investigate and interpret the phytogeography of Early
Devonian floral assemblages. For convenience, we use the term *differentiation* to refer to the phytogeographic differentiation of floral
assemblages. Most Early Devonian paleogeographic reconstructions
show emergent land surfaces extending from 60°N paleolatitude to the
Devonian south pole (Barrett, this volume; Scotese et al., 1979; Smith,
Hurley, and Briden, 1981). These land areas, most of which have Early
Devonian (Gedinnian through Emsian) floral assemblages, must have
experienced a wide range of climatic conditions. Because of this, we
expected Early Devonian floral assemblages to show differentiation.

Such differentiation has already been observed for the last half of
the Early Devonian (Edwards, 1973). In this study, we undertake a
more detailed analysis of the phytogeography of this period using the
powerful technique of multivariate statistical analysis. Only for the last
half of the Early Devonian is there sufficient reported data to warrant
the use of such techniques.

We also present the first analysis of the phytogeography of the
earliest Devonian (Gedinnian and early Siegenian). Here we employ
the more traditional method, identification of regionally shared endemic
taxa, since the scanty data render sophisticated statistical analysis
unnecessary.

The results of our analysis are a more detailed understanding of the

129

phytogeography of the last half of the Early Devonian and the first indication of differentiation in the earliest Devonian. We evaluated these results using the paleogeographic reconstructions and paleoclimatic models developed by Barrett (this volume). We find that these models explain the observed differentiation fairly well. Using these reconstructions and the analyses of differentiation possible with multivariate statistics, we lay the groundwork for studies of the climatic adaptions and dispersal of plants in the early Paleozoic.

This study is one in a series investigating the floral differentiation of the Paleozoic and Mesozoic and its correlation to climate and geography. It is an extension of earlier work by Gierlowski (1978) and Ziegler et al. (1981) on Early Devonian phytogeography. This study incorporates a larger data set than either of those previous studies and improved analytical techniques.

Previous Work

Originally, it was believed that the Early Devonian flora was cosmopolitan (Arber, 1921), although Andrews (1961) suggested that this uniformity had been overemphasized. Petrosyan (1967) distinguished three Early Devonian phytogeographic units within Eurasia. In the first global study, Edwards (1973) recognized differentiation between assemblages in Gondwana and those in the northern paleolatitudes and was able to distinguish further separation within the latter. Both Petrosyan and Edwards relied on the identification of regionally shared endemic taxa to distinguish phytogeographic units. Using multivariate statistical methods, Gierlowski (1978) and Ziegler et al. (1981) confirmed and improved upon the results of Edwards.

The three phytogeographic units distinguished by Petrosyan (1967) were Europe, Kazakhstan, and Tunguska (Siberia). Early Devonian floral assemblages from the European portion of the U.S.S.R. are very similar to those of North America. The Early Devonian floral assemblages of Tunguska (Siberia) contain many endemic genera, including *Eldychemia*, *Bjertdagia*, *Hoegophyton*, and *Crisophyton*. However, Tunguskan assemblages also contain genera common in Europe and North America: *Eogaspesia*, *Drepanophycus*, *Psilophyton*, and *Taeniocrada*. The regional endemic, *Liadesmophyton*, characterizes Kazakhstanian assemblages.

Edwards (1973) recognized differences between Early Devonian floral assemblages from Gondwana and those from more northern paleolatitudes. The regionally shared endemic genus *Haplostigma* characterizes Gondwanian assemblages; northern-hemisphere genera such as *Psilophyton*, *Taeniocrada*, and *Zosterophyllum* seldom occur in these assemblages. Edwards recognized the same northern-hemisphere units as did Petrosyan (1967).

The use of multivariate statistical techniques verified that differentiation existed in the Early Devonian. Using cluster analysis, Gierlowski (1978) recognized differences between Gondwanian assemblages and those from more northern paleolatitudes. Within the Gondwanian group, Australian assemblages showed some differentiation from South American, African, and Antarctic assemblages; and one anomalous assemblage from Argentina grouped with northern-paleolatitude assemblages. Among the assemblages from more northern paleolatitudes, those from Siberia, Russia, and New Brunswick appeared to be anomalous. Gierlowski (1978) attributed this latter distinctness to the high diversity of assemblages from the U.S.S.R. and to recent work on the New Brunswick assemblages. This high diversity and the newly revised taxonomy caused these assemblages to appear different from others.

Using gradient analysis, Ziegler et al. (1981) described three Early Devonian phytogeographic units: a Gondwanian unit consisting of floral assemblages from Argentina, the Falklands, Brazil, Bolivia, South Africa, and Antarctica; an Equatorial unit consisting of assemblages from eastern North America, Europe, Kazakhstan, Siberia, China, and Australia; and a Western North American unit consisting of assemblages from Alaska, western North America, Spitsbergen, and Bathurst Island.

Within the Equatorial unit, Ziegler et al. (1981) recognized three intergrading subunits: an Appalachian subunit consisting of assemblages from eastern North America, the Acadian terrain (Scotland and the Canadian Maritimes), as well as Portugal and Libya; a Western European subunit consisting of western European assemblages and one Acadian assemblage (Southern Wales); and an Eastern European subunit consisting of assemblages from eastern Europe, China, Kazakhstan, Siberia, and Australia.

Methods

In our analysis of Early Devonian phytogeography, we follow the procedures described in Raymond, Parker, and Parrish (this volume), which outline three considerations crucial to any phytogeographic analysis. First, one must identify differentiation among floral assemblages, which we do using multivariate statistics. We apply this technique to a larger data set than has been the case in past work, allowing a more comprehensive and detailed study of differentiation. The section below on data analysis briefly describes our analytical method.

Second, one must separate phytogeographic differentiation from differentiation due to other factors. For instance, disparity in habitat, stratigraphic age, and the nationality of the worker describing the assemblage can introduce apparent phytogeographic differentiation. We have collected our data base so as to eliminate as many of these factors as possible. A brief discussion of these factors appears in the section on data collection in this paper; a fuller discussion appears in Raymond, Parker, and Parrish (this volume). In addition, our method of data analysis allows us to identify false phytogeographic differentiation in the results *a posteriori*.

Third, one must interpret the observed differentiation in light of paleogeographic reconstructions and paleoclimatic models. Both paleoclimate and paleogeography are important in considering the results of an analysis of ancient phytogeography because of convergent evolution in response to climate. We discuss this topic next.

Interpretation of Ancient Phytogeographic Analyses

Although regional distributions of plant taxa form the data base for both modern and ancient phytogeography, the interpretation of ancient analyses differs from that of modern ones. To the extent that the systematics of modern taxa (based on reproductive characters) reflects phylogeny, modern phytogeography separates regions on the basis of phylogeny. In contrast, Paleozoic plant compression taxa are based largely on vegetative morphology; reproductive characters are used only when these are known. Thus, these taxa are determined by both morphology and phylogeny. Ancient phytogeography may therefore separate regions on the basis of nonphylogenetic factors that control morphology as well as phylogeny.

Paleoclimate may be one of these factors. In modern plants climate influences morphology, as indicated by the field of ecophysiology founded by Schimper (1903) and continued by workers such as Bailey and Sinnott (1915, 1916) and Walter (1973). If the relationship between climate and plant morphology obtained in the Paleozoic, then ancient phytogeographic units reflect a mixture of paleoclimatic and phylogenetic factors. For this reason, paleoclimate models are essential for the interpretation of ancient phytogeography.

We assume that morphological separation implies the presence of some genetic separation between regions and that Paleozoic phytogeography reflects biogeographic differentiation in the modern sense. However, the units delineated represent a conservative view of the amount of differentiation present. If climate caused convergent evolution among Paleozoic plants, each ancient unit may include more than one "phylogenetic" biogeographic unit.

Figure 6.1 illustrates one possible relationship between modern

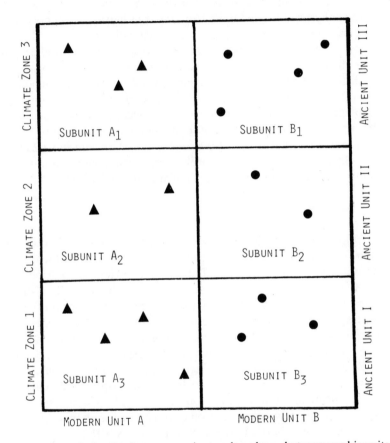

Fig. 6.1. The relationship between ancient and modern phytogeographic units. Two hypothetical modern units (A and B) are shown separated by a vertical line. Each modern unit contains three subunits (A_1, A_2, A_3 and B_1, B_2, B_3), separated by horizontal lines, which correspond to climatic zones 1 through 3. The assemblages of modern unit A (triangles) are more closely related phylogenetically to one another than to the assemblages of modern unit B (dots). Ancient phytogeographic units correspond to the three climatic zones and each contains assemblages from modern unit A and modern unit B. Because of the criteria used to delineate modern and ancient units differ, the assemblages within ancient units are not more closely related to one another than to assemblages from another unit.

and ancient phytogeographic units. In it, two hypothetical modern phytogeographic units (separated by a vertical line) contain three phylogenetically related but morphologically distinct subunits corresponding to different climatic zones. Four hundred million years in the future, a phytogeographic analysis of the same area might yield three "ancient" phytogeographic units (separated in figure 6.1 by horizontal lines) corresponding to the three original climatic zones.

The relationship of individual assemblages within modern and ancient phytogeographic units differs. In modern unit A, the morphologically distinct assemblages of each subunit are more closely related to one another than to any assemblage in modern unit B. Although the assemblages of ancient unit 1 are morphologically similar, the assemblages that make up ancient unit 1 are not necessarily more closely related to one another than to the assemblages of ancient units 2 and 3. Either phylogenetic factors or morphological convergence could cause the morphological similarity of assemblages in ancient unit 1.

Data Collection

Criteria for the Inclusion of Assemblages. Raymond, Parker, and Parrish (this volume) provide an extensive discussion of the factors that act to obscure phytogeographic differentiation during time periods with gentle latitudinal temperature gradients and low diversity. This discussion also applies to Early Devonian times. One of the major problems is separating the differentiation caused by disparity in age from that caused by paleoclimate and to allow for the possibility of differentiation due to age or habitat in the interpretation of our results.

We avoided using lists contained in regional compilations, unless they were accompanied by detailed locality descriptions. Because we could not locate all of the original sources used in Petrosyan (1967), our study excludes a few of the regional endemics listed therein. Floral assemblages from Siberia, Podolia (European U.S.S.R.), and Mongolia (Ananiev and Krasnov, 1962; Ischenko, 1965, 1974; and Chetverikova et al., 1971) represent possible exceptions to the rule of using assemblages from individual formations or members. Based on locality descriptions, we suspect that these floral assemblages come from limited geographic areas but possibly include more than one outcrop or locality.

Criteria for the Inclusion of Taxa. We chose the genus as the taxonomic unit for this study. This eliminates some of the problems posed by the limited stratigraphic and paleogeographic distributions of fossil species, as well as by incorrect species-level taxonomy. It also enables us to use lists of assemblages identified at the generic but not the specific level.

We attempted to limit the genera used in this analysis to those listed by Chaloner and Sheerin (1979) or Banks (1980); however, we included endemic taxa that were not listed in these references but that appeared in more than one floral assemblage. This egalitarian approach to gathering data has two disadvantages. (1) Erroneously identified genera may be included in the data. To minimize this possiblity, we excluded genera designated "cf."; (2) Endemic genera might reflect national taxonomic practices rather than phytogeographic differentiation, causing the nationality of the worker to determine the phytogeographic affiliation of the floral assemblage. However, the results presented here suggest little or no national bias.

Because both morphology and phylogeny determine fossil taxa, all taxa used herein potentially contained unrelated forms. As previously discussed, if paleoclimate influenced the morphology of these unrelated forms, then the ancient phytogeographic units described in this chapter may reflect regions with different paleoclimates as well as with phylogenetically distinct plant fossils (see figure 6.1). However, we did exclude two Early Devonian genera, *Hostinella* and *Thursophyton*, which refer respectively to plant stems without sporangia, leaves, or spines and to plant stems with spines but without sporangia. These genera possess extremely nondiagnostic morphologies. Their presence may indicate poorly preserved fossils in which diagnostic features have been destroyed rather than the morphology of the original community.

Our criteria for the inclusion of genera in this analysis differ somewhat from those of Raymond, Parker, and Parrish (this volume) in that we included genera identified on the basis of attached reproductive organs. Raymond, Parker, and Parrish excluded reproductive genera from their analyses because reproductive organs (seeds, cones, megaspores, and sporophylls) and vegetative organs (leaves and bark) often become sorted into separate compression floral assemblages during deposition (Spicer, 1980; Krasilov, 1975). We here include genera displaying attached reproductive organs because reproductive characters (sporangial form and attachment) determine many Early Devonian floral taxa. We assume that attached sporangia did not greatly influence the sorting of Early Devonian compression fossils during deposition.

Data Analysis

Treatment of Data. In contrast to previous multivariate studies (Gierlowski, 1978; Ziegler et al., 1981), which combined the fossil plant assemblages known from a geographic region into a single "average" list, we kept the floral assemblage lists separate during the multivariate statistical analysis. This has three advantages:

1. We can differentiate between rare and common endemic taxa within a region;
2. We decrease the effect that anomalous floral assemblages have on the floristic composition of the region as a whole;
3. We decrease the distortion that may result when a multivariate statistical analysis uses samples of differing sizes.

The disadvantage of using combined floral lists has been revealed in previous work. Ziegler et al. (1981) and Gierlowski (1978) used both samples from single bedding planes and samples that represented regional compilations. In cluster and ordination analyses, floral assemblages group together because they share taxa. If assemblage lists are combined so that regionally shared endemic taxa appear only once in the data set, then these taxa, which are phytogeographically important, can contribute little information to the analysis. An analysis that ignores regionally shared endemic taxa will reflect only the differentiation caused by the gradational distribution of widely occurring taxa.

Raymond, Parker, and Parrish (this volume) discuss how combining floral assemblage lists may also obscure phytogeographic differences if assemblages from possibly displaced terrains (such as Acadia) are lumped with those from nearby stable cratons. Gierlowski (1978) and Ziegler et al. (1981) combined floral assemblages into regional lists on the basis of detailed paleogeographic reconstructions and did not lump assemblages from cratons and displaced terrains.

However, the regions used by Gierlowski (1978) and Ziegler et al. (1981) differed greatly in areal extent. Some included entire paleocontinents (for example, Kazakhstan and Siberia), whereas others included only small regions or single bedding-plane assemblages (for example, France, Belgium, Bathurst Island, and Wyoming). The difficulty of obtaining well-documented assemblages from single bedding planes from countries such as the U.S.S.R. caused this disparity among the size of the regions used in these previous studies. Such a disparity may distort the results in at least two different ways: differences in diversity (genus richness) may distort the results of a multivariate analysis; and fine-scale phytogeographic differentiation may be obfuscated.

Diversity generally increases with the size of the region sampled (Pianka, 1974; MacArthur and Wilson, 1967); thus floral assemblages from large regions contain more genera than assemblages from single bedding planes. For example, the region list for Siberia contained thirty genera; the most diverse single bedding-plane assemblage (Gaspé) contained nine genera, while most contain between one and six (Gierlowski, 1978; see also Appendix, below). The use of assemblages that vary widely in diversity may distort the results of a multivariate analysis (Schmachtenberg, 1983). In addition, the use of combined floral lists prevented Ziegler et al. (1981) from observing fine-scale phytogeographic differentiation within the poorly sampled area (Siberia, China, and Kazakhstan).

In order to produce the multivariate statistical analyses in this study, we arranged the data into a matrix with 46 assemblages (samples) and 23 genera (variables). The data are in presence-absence form (see table 6A. 2 in the Appendix); a genus is "present" at a given locality if it occurred in the floral list from that locality and "absent" if it did not appear in the floral list. We would have preferred to use data that reflected the relative abundance of the genera in each floral assemblage, but this information was not available in the published reports of Early Devonian floral assemblages.

Multivariate Statistical Techniques. We used two multivariate statistical techniques to investigate phytogeographic differentiation among floral assemblages: cluster analysis and polar ordination analysis. Given m genera, all of the variation in n assemblages could be explained by plotting the occurrence and variation of those n assemblages in m dimensions. If two genera (variables) are used to characterize a number of assemblages, two dimensions are sufficient to describe all the variation in the samples. However, larger data sets with many variables are difficult to comprehend; in this study we would need 23 dimensions (one per genus) to encompass all the variation in the assemblages. Multivariate statistical techniques reduce the dimensionality of data sets to a number of dimensions large enough to explain most of the variation present but small enough to comprehend.

Cluster Analysis   Cluster analysis sorts assemblages into groups based on generic similarity. Floral assemblages that share many genera have high similarity and belong to the same cluster, whereas those that share few genera will belong to different clusters. For a discussion of the applications and variations on the basic method of cluster analysis, see Sneath and Sokal (1973). For this study, we used a Fortran computer program, CLAP, written by J. J. Sepkoski, Jr.

The result of a cluster analysis is a dendrogram in which the level of clustering indicates the relative similarity between two samples. For instance, in the results of our cluster analysis (figure 6.3), localities 27 and 31 cluster together near the level of 0.8. These localities are less similar to one another than are localities 30 and 32, which cluster together at the level of 0.99.

Cluster analysis begins with the computation of a similarity matrix. We used the Otsuka similarity index, which is less sensitive to variations in sample size than other similarity indices (Cheetham and Hazel, 1969). The formula for the Otsuka similarity index is

$$S_O = \frac{a}{(n_1 \cdot n_2)},$$

where a = the number of genera shared by assemblages 1 and 2; n_1 = the number of genera in assemblage 1; and n_2 = the number of genera in assemblage 2 (Cheetham and Hazel, 1969). Clustering starts with the most similar sample and proceeds iteratively to the least similar sample, forming a dendrogram similar to the one shown in figure 6.3.

Strengths of Cluster Analysis   Cluster analysis yields easily interpretable results; in addition, the technique is easy and inexpensive to perform on most computors. Cluster analysis reduces dimensionality while preserving the relationships among the samples within a cluster. However, cluster analysis may distort the relationships among clusters and among samples belonging to different clusters. In other words, we expect cluster analysis to identify phytogeographic units but realize that it might not reflect gradational relationships among these units.

Polar Ordination   Polar ordination is one of a family of techniques known as gradient analysis, introduced by Bray and Curtis (1957) to investigate distributions of modern plants. For discussions of the applications and variations of the basic method of polar ordination, see Beals (1965), Whittaker (1978), Orloci (1978), and Cisne and Rabe (1978). To do the polar ordination presented in this study, we used a Fortran computer program, POLAR II, written by J. J. Sepkoski, Jr., and J. Sharry.

The results of polar ordination are a series of two-dimensional plots that map the floral similarity of assemblages. The distance between assemblages on the plot reflects the relative floral similarity of those assemblages. For example, in the results of the polar ordination analysis presented here (figure 6.4), locality E-1 plots close to locality E-13 and far from locality Aus-3; locality E-1 shares many genera with locality E-13 and few with locality Aus-3.

Like cluster analysis, polar ordination begins with the calculation of a similarity matrix. For this analysis, we used two similarity indices to calculate this similarity matrix, the Otsuka index (defined above) and the Dice index:

$$S_D = \frac{2a}{(n_1 + n_2)},$$

where a = the number of genera shared by assemblages 1 and 2; n_1 = the number of genera in assemblage 1; and n_2 = the number of genera in assemblage 2 (Cheetham and Hazel, 1969). Both indices gave similar polar ordination patterns, but the Dice similarity index explained more of the variation in the data set; we have reported only the Dice index results.

In order to produce a polar ordination diagram, the similarity between two points is converted into "dissimilarity" (one minus the similarity value), and dissimilarity is used as a measure of distance between assemblages. The assemblage that shows the greatest variance in dissimilarity relative to all other assemblages forms the endpoint of the first ordination axis. The assemblage that is most dissimilar to the first endpoint forms the second endpoint of that axis. The positions of all other assemblages are projected onto this axis so that the distance between assemblages on the axis is proportional to their original dissimilarity. Endpoints for axes 2 and 3 are chosen in a similar manner, starting with the non-endpoint assemblage that shows the greatest variance in dissimilarity relative to the other assemblages. Subsequent axes are constrained to be roughly perpendicular to, and thus uncorrelated with, the first axis. The results of POLAR II provide an indication of the amount of variation in the similarity matrix that each axis explains.

POLAR II also provides an indication of the variables (genera) that determine the position of each sample on the axes. The relationship between the genera (variables) and the ordination axes is given by the ordinated variable positions. The positions are calculated for each genus by summing the ordinated position of each assemblage containing the genus and dividing by the total number of assemblages. Genera that ordinate high on an axis characterize assemblages that ordinated high on that axis.

Strengths of Polar Ordination Polar ordination reveals gradational relationships among floral assemblages, if such exist in the data. Polar ordination complements cluster analysis. Used in conjunction with that technique, polar ordination can show the relationships between clusters and also the relationships of samples from different clusters, both of which may be distorted in the results of a cluster analysis. Like cluster analysis, polar ordination yields easily interpretable results. In addition, it can be done on most computers or, if the number of samples is small, with a hand calculator.

Results

Middle Siegenian–Emsian Phytogeography

The data for this analysis consisted of information about the presence or absence of 23 floral genera within 46 compression floral assemblages; the data appear in table 6A.2 of the Appendix. Figure 6.2 shows the localities of the middle Siegenian and Emsian floral assemblages used in this study, plotted on the Emsian paleogeographic reconstruction

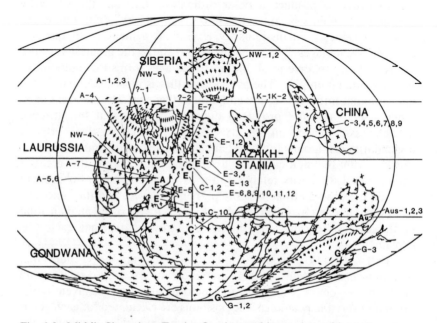

Fig. 6.2. Middle Siegenian–Emsian floral assemblages. Assemblage abbreviations indicate their position on the cluster analysis dendrogram (see fig. 6.3). A list of the assemblage localities, their present-day latitudes and longitudes, and their abbreviations appears in table 6A.1 of the Appendix. Assemblages marked N clustered with Siberian assemblages; A, with American; E, with European; K, with Kazakhstanian; C, with Chinese; Au, with Australian; G, with Gondwanian. The following assemblages lie very close to the suture between northern and southern Europe and may actually be on the southern European plate, which is shaded grey in this figure: C-1 (Germany-1); C-2 (Germany-2); E-6 (Houyoux R., Belgium); E-8 (Meuse R., Belgium); E-9 (Estinnes au Mont, Belgium); E-10 (Carrière à Dave, Belgium); E-11 (Bruay, Belgium); and E-12 (Matringham, Belgium).

proposed by Barrett (this volume). The paleocontinent on which an assemblage grew and the paleolatitude at which it grew provide information that can be used to interpret paleophytogeographic patterns. For convenience, we will use the following terms: *equatorial* for the paleolatitudes between 15°N and 15°S; *low latitude* for those between 15°N and 30°N or 15°S and 30°S; *middle latitude* for those between 30°N and 50°N or 30°S and 50°S; and *high latitude* for those between 50°N and 90°N or 50°S and 90°S.

Figure 6.3 shows the results of the cluster analysis of all middle Siegenian–Emsian floral assemblages. Cluster analysis reveals three major groups of floral assemblages at the 0.3 level of the Otsuka simila-

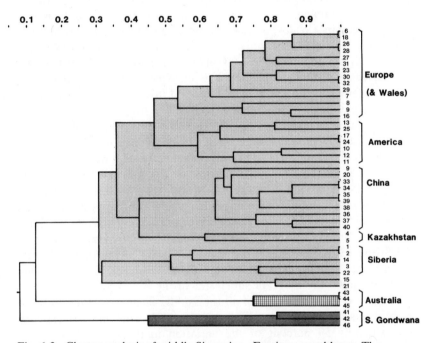

Fig. 6.3. Cluster analysis of middle Siegenian–Emsian assemblages. The dendrogram shows three biogeographic units clustering together at the 0.3 level: an equatorial and low-latitude unit (stippled); an Australian unit (grid pattern); and a Southern Gondwanian unit (diagonal lines). Within the Equatorial and low-latitude unit, cluster analysis revealed five subunits, clustering together at the 0.5 level: Europe plus Wales; America, which includes all Acadian localities except Wales; China, which includes northern Africa and two assemblages from Germany; and Siberia, which includes assemblages from northwestern North America.

rity index; a large cluster consisting of floral assemblages from Europe, China, Kazakhstan, Spitsbergen, North America, and Siberia (stippled area); a small cluster of floral assemblages from Australia (grid); and a small cluster consisting of floral assemblages from Antarctica and South Africa (diagonal lines). The Australian floral assemblages lay near the Emsian paleolatitude of 30°S; the floral assemblages from Antarctica and South Africa lay in southern Gondwana in the south high latitudes (see fig. 6.2).

Assemblages in the largest cluster of figure 6.3 plot between the paleolatitudes of 60°N and 40°S, the paleocontinents of Laurussia, Siberia, Gondwana, Kazakhstan, and China. Some internal structure appears in this large cluster at similarity levels greater than 0.5. Most of the floral assemblages from Europe and one assemblage from the Acadian terrain (Wales) cluster at the 0.53 level. We labeled this subcluster "Europe" on figure 6.3. The other assemblages from the Acadian terrain (Scotland and Oxford), and those from New England and eastern Canada (Maine, New Brunswick, Ontario, and Gaspé) cluster together at the 0.6 level. We labeled this subcluster "America." All the assemblages from China, as well as two from Germany and one from the Sahara Desert, cluster together at the 0.64 level. We labeled this subcluster "China." The two assemblages from Kazakhstan cluster together at the 0.61 level. This subcluster, labeled "Kazakhstan" on figure 6.3, joins the Chinese subcluster at the 0.42 level and is more closely related to the Chinese subcluster than to the Acadian or European subclusters. Assemblages from Siberia, the western United States, and Spitsbergen also cluster above the 0.5 level. We labeled this subcluster "Siberia." Finally, the assemblage from the Harz Mountains of Europe (assemblage 21) and Bathurst Island in northern Canada (assemblage 15) cluster together above the 0.5 level. This unlabeled subcluster joins the Siberian subcluster at the 0.31 level.

Figure 6.4 shows a plot of all the middle Siegenian–Emsian floral assemblages on polar ordination axes 1 and 2. This analysis explains approximately 59 percent of the variation in the original similarity matrix on three axes and reveals the relationships among the clusters described in figure 6.3. Polar ordination axis 1 separates Australian assemblages, which formed a cluster at the 0.3 level, from the Kazakhstanian assemblages, which form a subcluster at the 0.5 level. Axis 1 also separates the Kazakhstanian subcluster from the rest of the equatorial and low-latitude assemblages, which belong to the stippled cluster in figure 6.2. This axis accounts for approximately 31 percent of the variation in the original similarity matrix. Polar ordination axis 2 separates the southern Gondwanian assemblages, which formed a

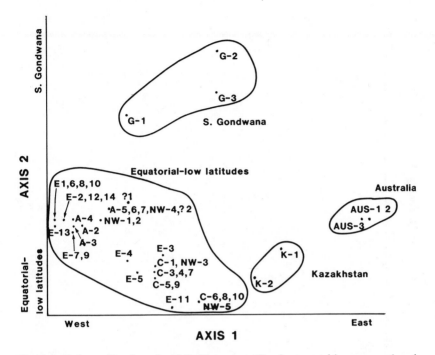

Fig. 6.4. Polar ordination of middle Siegenian–Emsian assemblages, axes 1 and
2. Assemblage abbreviations indicate their position on the cluster
analysis dendrogram (fig. 6.3); a list of these abbreviations appears in
table 6A.1 of the Appendix. Assemblages from southern Gondwana,
Australia, Kazakhstan, and the rest of the equatorial and low latitudes
all plot in separate areas on the ordination diagram.

cluster at the 0.3 level, from all the equatorial and low-latitude
assemblages.

Biogeographically Significant Genera. Figure 6.5 indicates the
genera that are important in determining the position of assemblages in
the ordination analysis. This plot of ordinated variable positions shows
the genera plotted on the same axes as are the assemblages in figure 6.4.
The position of each genus is calculated by summing the ordinated
position of each assemblage containing the genus, dividing by the total
number of assemblages in which the genus occurs, and plotting the
genus position on the polar ordination axes. The most biogeographically
significant genera plot near the endpoints of the axes.

The genera that characterize Australian assemblages (*Yarravia*,

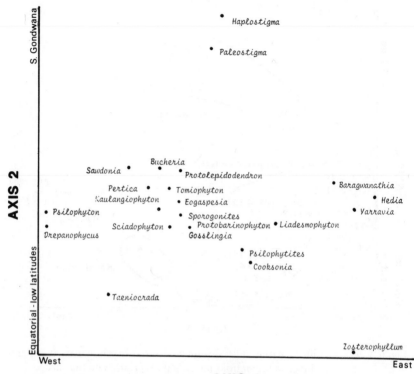

Fig. 6.5. Plot of ordinated variable positions, axes 1 and 2. This plot shows the
biogeographically significant genera in the ordination analysis of
middle Siegenian–Emsian floral assemblages, plotted on the same
axes shown in figure 6.4. The most biogeographically significant genera
appear at the endpoints of the axes.

Hedia, and *Baragwanathia*) plot high on axis 1. In figure 6.4, this axis
separates Australia from other equatorial and low-latitude assemblages.
The Kazakhstanian assemblages plot close to the Australian assemblages because *Hedia* occurs in both regions. The other genus that
characterizes Kazakhstanian assemblages, *Liadesmophyton*, also plots
high on axis 1 in figure 6.5. The genera that characterize equatorial and
low-latitude assemblages (*Psilophyton and Drepanophycus*) plot low on
axis 1 in figure 6.5, in the ordination position of equatorial and
low-latitude assemblages in figure 6.4.

The genera that characterize southern Gondwanian assemblages
(*Haplostigma* and *Paleostigma*) plot high on axis 2 in figure 6.5 and

occupy of the ordination position of southern Gondwanian assemblages in figure 6.4. The genera that plot low on axis 2 in figure 6.5 (*Zostero-phyllum* and *Taeniocrada*) characterize many of the equatorial and low-latitude floral assemblages. These genera plot in the ordination position of assemblages from equatorial and low latitudes, including Kazakhstan.

Differentiation in Equatorial and Low Latitudes. In the cluster analysis of middle Siegenian–Emsian assemblages (figure 6.3), we observed differentiation within the large equatorial and low-latitude cluster. However, the polar ordination (figure 6.4) does not reveal much differentiation within these assemblages; floral assemblages that form separate subclusters on figure 6.3 intermingle in figure 6.4. For instance, on figure 6.4 Podolia (assemblage 8), which belongs in the European subcluster, plots closest to New Brunswick (assemblage 10), which belongs to the American subcluster.

Because we wanted to investigate the possibility of differentiation among these equatorial and low-latitude assemblages, we completed a cluster analysis and a polar ordination using only the equatorial and low-latitude assemblages (all of which clustered together at the 0.3 level in the stippled cluster of figure 6.3). This analysis yielded results identical to that part of the cluster analysis of all middle Siegenian–Emsian assemblages.

Figures 6.6 through 6.10 show the results of the polar ordination analysis of the equatorial and low-latitude floral assemblages. This analysis, which used the Dice similarity index, explained approximately 63 percent of the variation in the original similarity matrix on three axes. Figure 6.6 is a plot of floral assemblages on polar ordination axes 1 and 2. Axis 1, which accounted for approximately 35 percent of the variation in the original similarity matrix, displays a gradient between eastern-equatorial and low-latitude assemblages (China and Kazakhstan) and western-equatorial and low-latitude assemblages (Scotland, Ontario, and Bathurst Island). The assemblages included in the Kazakhstanian and Chinese subclusters plot low on axis 1; the assemblages from the Acadian and European subclusters are intermixed along axis 1. Axis 2, which accounts for approximately 25 percent of the variation in the original similarity matrix, separates the Kazakhstanian assemblages from all other tropical and subtropical assemblages.

Figure 6.7 is a plot of floral assemblages on polar ordination axes 1 and 3. Polar ordination axis 3, which returned approximately 17 percent of the variation in the original similarity matrix, separates the floral assemblages from Siberia, Spitsbergen, and Wyoming from the other

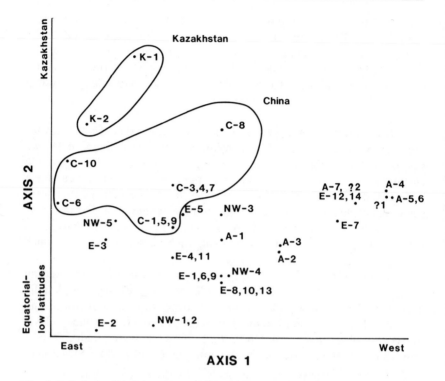

Fig. 6.6. Polar ordination of equatorial and low-latitude middle
Siegenian–Emsian assemblages, axes 1 and 2. Assemblages from both
the Kazakhstanian and the Chinese subunits of figure 6.3 plot in
separate areas of the ordination diagram. Assemblages from the
European, the Siberian, and the American subunits of figure 6.3 are
intermixed along axis 1.

equatorial and low-latitude floral assemblages. The floral assemblages
separated on axis 3 belong to the Siberian subcluster of figure 6.3.

Figure 6.8, the plot of equatorial and low-latitude floral assemblages on polar ordination axes 2 and 3, summarizes the relationships
among floral assemblages as shown by cluster analysis and polar
ordination. On this figure, we outlined each of the subclusters from the
stippled cluster of figure 6.3. The Siberian, Kazakhstanian, and Chinese
subclusters do not overlap any other subclusters. However, the American and European subclusters, although separable, plot extremely close
together. The results of the polar ordination analysis considered in
conjunction with cluster analysis provide the basis for separating these
groups.

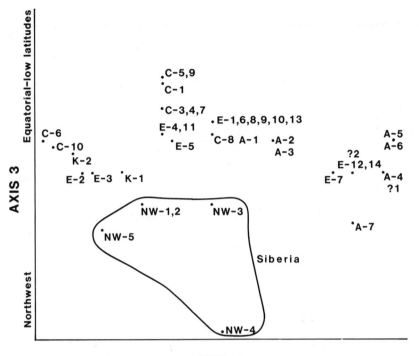

Fig. 6.7. Polar ordination of equatorial and low-latitude middle
Siegenian–Emsian assemblages, axes 1 and 3. Assemblages from the
Siberian subunit of figure 6.3 plot in a separate area of the ordination
diagram.

Figure 6.8 suggests a northwest-southeast gradient linking the
Siberian subcluster to the American subcluster, the American subcluster to the European subcluster, the European subcluster to the
Chinese subcluster, and the Chinese subcluster to the Kazakhstanian
subcluster.

Biogeographically Significant Genera of Equatorial and Low Latitudes. Figures 6.9 and 6.10 show the plots of ordinated variable position
for the equatorial and low-latitude analysis. These plots indicate the
genera important in determining the ordination position of floral
assemblages.

Figure 6.9 is a plot of vegetative genera that occur in equatorial and
low-latitude assemblages on polar ordination axes 1 and 2. In the

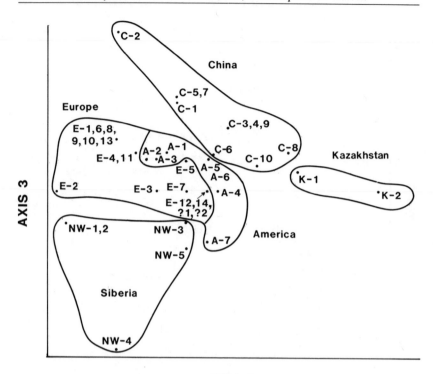

AXIS 2

Fig. 6.8. Polar ordination of equatorial and low-latitude middle
Siegenian–Emsian assemblages, axes 2 and 3. Assemblages from the
five named subclusters of figure 6.3 (Siberia, Europe, America, China,
and Kazakhstan) plot in separate areas of the ordination diagram,
along a northwest-to-southeast gradient that links Siberia to Europe
and America, Europe and America to China, and China to
Kazakhstan.

ordination analysis of figures 6.4 and 6.5, axis 1 displayed a gradient
between eastern-equatorial and low-latitude assemblages and western-
equatorial and low-latitude assemblages. Figure 6.9 shows that *Taeniocrada*, which characterizes eastern-equatorial and low-latitude assemblages, plots low on axis 1 and that *Sawdonia* and *Drepanophycus*,
which occur in western-equatorial and low-latitude assemblages, plot
high on axis 1.

  Sawdonia was distinguished from *Psilophyton* relatively recently
(Hueber, 1971b). Because of this, recent descriptions of floral assem-

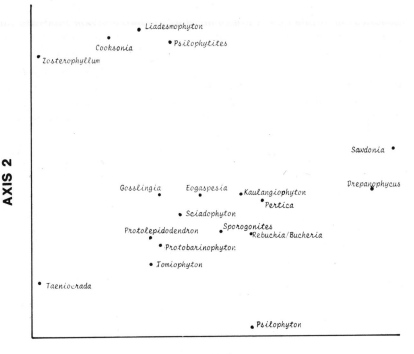

AXIS 1

Fig. 6.9. Equatorial and low-latitude ordinated variable positions, axes 1 and 2.
This plot shows the biogeographically significant genera in the
ordination analysis of equatorial and low-latitude assemblages plotted
on the same axes as in figure 6.6. Genera responsible for the
separation of Kazakhstan and China plot high on axis 2.

blages from North America and western Europe might be more likely to
include *Sawdonia* than might descriptions of floral assemblages from
Eastern Europe and the U.S.S.R. Thus, the east-to-west differentiation
along axis 1 in the polar ordination of equatorial and low-latitude
assemblages may relate to regional taxonomic practices or the date of
description of the published floral assemblages. However, because
recently described floral assemblages from China do not include *Sawdonia*, we suspect that the observed east-to-west differentiation reflects
phytogeographic differentiation.

In the ordination of equatorial and low-latitude floral assemblages
(fig. 6.6), axis 2 differentiated the floral assemblages from Kazakhstan

from other equatorial and low-latitude assemblages. In figure 6.9, the genera *Zosterophyllum*, *Cooksonia*, and *Liadesmophyton* plot high on axis 2, near the ordination position of the assemblages from Kazakhstan. The genus *Liadesmophyton* occurs only in Kazakhstan; *Cooksonia* occurs in Kazakhstan, China, the Sahara Desert, and Wales during the middle Siegenian and Emsian. The genus *Zosterophyllum* occurs in all floral assemblages from China and Kazakhstan, as well as in floral assemblages from Europe and North America on the paleocontinent of Laurussia and from Siberia.

The genera *Taeniocrada* and *Psilophyton* plot low on axis 2. The genus *Taeniocrada* characterizes western-equatorial and low-latitude

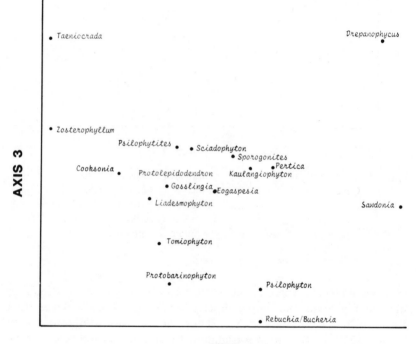

AXIS 1

Fig. 6.10. Equatorial and low-latitude ordinated variable positions, axes 1 and 3. This plot shows the biogeographically significant genera in the ordination analysis of equatorial and low-latitude assemblages plotted on the same axes as in figure 6.7. Genera responsible for the separation of Siberia and northwestern North America plot low on axis 3.

assemblages. The genus *Psilophyton* occurs in most equatorial and low-latitude assemblages from Europe, the U.S.S.R., and North America; this genus does not occur in Kazakhstan or China.

In the ordination of equatorial and low-latitude floral assemblages (fig. 6.5), axis 3 differentiated floral assemblages from Wyoming, Spitsbergen, and Siberia from other equatorial and low-latitude floral assemblages. In figure 6.10, *Protobarinophyton/Barinophyton*, *Psilophyton*, and *Bucheria/Rebuchia* plot low on axis 3, in the position of assemblages from the Siberian subcluster.

In summary, the results of the polar ordination analysis and cluster analysis suggest the existence of three major phytogeographic units in the Early Devonian: a large equatorial and low-latitude unit that includes floral assemblages from China, Kazakhstan, Siberia, Laurussia, and northern Gondwana; a small southern, low-latitude unit that includes assemblages from Australia; and a small middle- to high-latitude unit that includes floral assemblages from South Africa and Antarctica.

Cluster analysis and polar ordination analysis of floral assemblages from the large equatorial and low-latitude unit reveal phytogeographic differentiation within that unit along a northwest-to-southeast gradient. We find evidence for five intergradational units: a northwestern Siberian unit, which includes the west coast of Laurussia; a western tropical unit consisting of eastern North American and Acadian assemblages; an eastern tropical unit consisting of one Acadian assemblage (Wales) and European assemblages; a Chinese unit consisting of Chinese assemblages, an assemblage from northern Gondwana, and two assemblages from Germany; and a Kazakhstanian unit, consisting only of assemblages from Kazakhstan.

Gedinnian–Early Siegenian Phytogeography

Very few floral assemblages from the Gedinnian and early Siegenian have been identified (see table 6.1). Although the number of described assemblages is too small for a multivariate statistical study, perusal of table 6.1 indicates the existence of two Gedinnian–early Siegenian phytogeographic units: an Australian unit characterized by *Yarravia* and *Baragwanathia* and a widespread "northern-hemisphere" unit characterized by *Zosterophyllum*, *Taeniocrada*, *Cooksonia*, and *Drepanophycus*. Figure 6.11 shows the location of these floral assemblages on the Emsian reconstruction proposed by Barrett (this volume); in the absence of a Gedinnian–early Siegenian reconstruction, the Emsian one is the best available.

Table 6.1   Gedinnian–Early Siegenian Floral Assemblages

Location	*Baragwanathia*	*Yarravia*	*Taeniocrada*	*Drepanophycus*	*Zosterophyllum*	*Psilophytites*	*Cooksonia*	*Sporogonites*	Reference
Yea, Australia 37°S, 145°E	1	1							Garratt, 1978
Matlock, Australia 38°S, 146°E	1	1							Garratt, 1978
Spitsbergen, Norway 80°N, 12°E			1	1					Høeg, 1942; Schweitzer, 1968
Nonceveaux, Belgium 50°N, 6°E			1	1					Leclercq, 1942
Wepion, Belgium 50°N, 5°E				1			1		Stockmans, 1940
Fofar, Scotland near 57°N, 2°W					1	1			Edwards, 1972
Pri-Balkash, Kazakhstan 47°N, 78°E					1	1	1		Chetverikova et al., 1971
Nurin, Kazakhstan 50°N, 75°E				1	1				Chetverikova et al., 1971
Chingiz Mt., Kazakhstan 48°N, 79°E					1	1			Chetverikova et al., 1971
Volcanic Suite, Kazakhstan unknown					1		1		Chetverikova et al., 1971
Xichuan, China near 25°N, 104°E						1	1	1	Li and Cai, 1978

Note: This table lists only those genera which occur in more than one Gedinnian–early Siegenian floral assemblage and omits *Sugambrophyton*, reported as a single occurrence in the Tonschiefer Group (Schmidt, 1954), and *Salopella*, which occurs only in the Welsh borderland (Edwards and Richardson, 1974). Edwards and Richardson (1974) accepted the Wepion, Belgium, assemblage as early Siegenian, although the presence of *Psilophyton princeps* may indicate a late Siegenian age (Banks, 1980; Chaloner and Sheerin, 1979).

Discussion

Middle Siegenian–Emsian Phytogeography

Cluster analysis and polar ordination analysis reveal three groups of middle Siegenian–Emsian floral assemblages that cluster at the 0.3 level. In the following discussion we will refer to these three groups as phytogeographic units. These units may reflect differences in climatical-

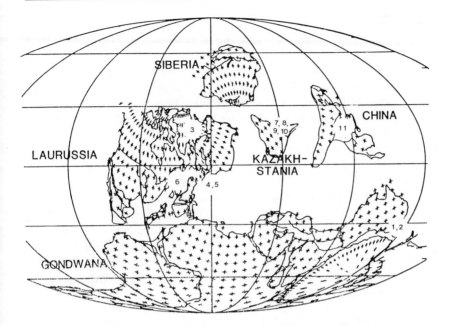

Fig. 6.11. Gedinnian and early Siegenian floral assemblages plotted on an Emsian reconstruction. A list of the assemblage localities and their present-day latitudes and longitudes appears in table 6A.1 of the text. Australian assemblages share no genera with other Gedinnian–early Siegenian floral assemblages and represent a separate Gedinnian–early Siegenian phytogeographic unit. Numbers refer to the following localities listed in table 6.1: 1 and 2, Yea and Matlock, Australia; 3, Spitsbergen; 4 and 5, Wepion and Nonceveaux, Belgium; 6, Fofar, Scotland; 7, 8, 9, and 10, Pri-Balkash, Nurin, Chingiz Mt., and Volcanic Suite, Kazakhstan, U.S.S.R.; 11, Xichuan, China. Localities 4 and 5 (Wepion and Nonceveaux, Belgium) lie very close to the suture between northern and southern Europe and may actually be on the southern European plate, shaded grey in this figure.

ly determined morphologies as much as they reflect phylogenetic divergence due to geographic separation. Because of this, we have purposely avoided using terms such as *kingdom*, *realm*, or *province*, which refer to biogeographic entities that result from the phylogenetic divergence of faunas or floras in geographically isolated regions. The largest of these units consists mostly of equatorial and low-latitude assemblages from the paleocontinents of Angara, China, Kazakhstan, and Laurussia but includes on floral assemblage from northern Gondwana. The other two units consist of southern low-latitude assemblages from Australia and southern middle- to high-latitude assemblages from Gondwana.

Cluster analysis of the equatorial and low-latitude unit revealed five subsidiary groups of assemblages from different phytogeographic regions that cluster together at the 0.5 level in figure 6.3. We refer to these groups as phytogeographic subunits; they are: the Kazakhstanian subunit; the Chinese subunit; the European subunit; the American subunit; and the Siberian subunit. Two equatorial and low-latitude floral assemblages, Bathurst Island and the Harz Mountains in Europe, clustered together at the 0.5 level of no apparent reason.

It is encouraging that cluster analysis placed most of the assemblages from the same paleogeographic region into the same cluster. At the 0.3 level, no assemblage was misplaced. At the 0.5 level, only five of the forty equatorial and low-latitude assemblages clustered with assemblages from a different paleogeographic region. Table 6.2 shows the affinity indicated by cluster analysis and the paleogeographic region of these assemblages.

Table 6.2   Early Devonian Assemblages Misplaced by Cluster Analysis

Assemblage	Number on Figure 6.3	Abbreviation	Paleogeographic Region	Floral Affinity Indicated by Cluster Analysis
Harz Mt.	21	?-2	European	Bathurst Is.
Bathurst Is.	15	?-1	Siberian	Harz Mt.
Germany-1	19	C-1	European	Chinese
Germany-2	20	C-2	European	Chinese
Wales	16	E-5	American	European

Polar ordination of the equatorial and low-latitude assemblages supports the results of cluster analysis and suggests that the five subunits lay along a paleogeographic gradient stretching from the Kazakhstanian and Chinese subunits in the southeast to the Siberian subunit in the northwest. While the Siberian, Kazakhstanian, and Chinese subunits can be clearly differentiated in polar ordination analysis, the European subunit and the American subunits intergrade to a large extent.

Differentiation Due to Age. Differences in the ages of assemblages do not appear to determine the units and subunits described in this study. The floral assemblages used belong to the middle Siegenian–Emsian *Psilophyton* zone of Banks (1980). The ages of assemblages from North America, Europe, the U.S.S.R., China, and Australia are fairly well constrained, although the assemblages from Gaspé, New Brunswick, and Trout Valley, Maine (Andrews et al., 1977), date from the Emsian-Eifelian boundary and may belong to the Middle Devonian rather than the Early Devonian. Since these assemblages cluster with other assemblages from North America and Acadia (Scotland and England), we assumed that their possibly aberrant age did not affect the analysis.

The floral assemblages from southern Gondwana (Plumstead, 1967) are poorly dated and may be Middle Devonian in age. However, these assemblages differ from Middle Devonian assemblages from the northern hemisphere (Edwards, 1973) and represent a separate phytogeographic unit, whether it is Early or Middle Devonian in age.

Comparison with Climatic Model. The Emsian paleoclimate model proposed by Barrett (this volume) explains the phytogeographic units defined in this study fairly well. Two of the three major units, the Australian unit and the equatorial and low-latitude unit, lay in the tropical-subtropical paleoclimate zones (see Barrett, this volume, fig. 5.12). The Australian unit lay near the boundary between southern subtropical humid and wet paleoclimatic zones. These assemblages would have received 100 cm or more rainfall annually and experienced a mildly seasonal climate with little freezing. The equatorial and low-latitude unit lay mostly in the tropical and northern subtropical wet paleoclimatic zones. The following equatorial and low-latitude assemblages are exceptions: Spitsbergen, tropical-subtropical humid; Bathurst Island, northern temperate dry; Minusinsk and the Lena River, boundary between tropical-subtropical humid and northern temperate subhumid; Sahara, shallow marine shelf closest to a tropical-subtropical humid land area. All equatorial and low-latitude assemblages except Bathurst Island would have received 100 cm or more rainfall annually and experienced a mildly seasonal climate with little freezing.

The third major unit, the southern Gondwanian unit, lay in southern temperate humid to subhumid paleoclimate zones. The southern Gondwanian assemblages would have received from 50 to 100 cm rainfall annually and experienced a cold seasonal climate.

If climatically influenced morphologies caused the differentiation of these major biogeographic units, we would have expected no difference between the Australian unit and the equatorial and low-latitude unit, since both belong to the same paleoclimate zone. Therefore, the differentiation of the Australian and the equatorial and low-latitude units may result from the phylogenetic divergence of Australian and other equatorial and low-latitude assemblages due to a dispersal barrier. The wide Tethys ocean and the Gondwanian desert (see Barrett, this volume, fig. 5.12) that separated Australia from the other equatorial and low-latitude assemblages could have constituted such a barrier to the dispersal of middle Siegenian–Emsian land plants. However, variations in paleoclimate not shown in the paleoclimatic model could also account for the differentiation between Australian and other equatorial and low-latitude assemblages. Finally, a mixture of paleoclimatic and phylogenetic factors could cause this differentiation. Better sampling of assemblages from equatorial and low-latitude paleoclimatic zones in northern Gondwana (northwestern Africa, India, and eastern Australia) may help resolve this question.

Within the equatorial and low-latitude unit, the Siberian subunit spans a wide variety of paleoclimatic zones (wet and humid tropical-subtropical to northern temperate subhumid). The western coast of Laurussia could have been cooler than equivalent paleolatitudes along the eastern coast of Laurussia, particularly if Tethyan circulation models apply to the Early Devonian (Barrett, this volume). Tethyan circulation could have resulted in Wyoming, a Siberian subunit assemblage at 0° paleolatitude on the west coast of Laurussia, being considerably cooler than Podolia, a European subunit assemblage at 2°S paleolatitude on the east coast of Laurussia. However, the wide paleoclimatic span of regions belonging to the Siberian subunit appears anomalous when compared with the paleoclimatic spans of regions assigned to other subunits: Chinese (tropical-subtropical wet and humid); Kazakhstanian (tropical-subtropical wet); European (tropical-subtropical wet); American (tropical-subtropical wet).

The assemblages from Siberia at 50°N paleolatitude contain many tropical and subtropical genera. Although we could not locate the specific floral assemblage, Petrosyan (1967) reported the presence of *Eogaspesia* in Siberia. Elsewhere (Podolia and Gaspé), this genus occurs within 5° of the paleoequator. Other genera with largely tropical and subtropical distributions that occur in Siberia are *Psilophyton*, *Taeniocrada*, and *Drepanophycus*. Floral links between the paleocontinent of Siberia and equatorial and low-latitude localities in the middle Siegenian and Emsian support the proposition of Barrett (this volume) that Siberia was situated further south in the Early Devonian than indicated by the reconstruction used in this study.

The results of this study are similar to those of Ziegler et al. (1981), who described three major Early Devonian phytogeographic units: (1) a Gondwanian unit, which corresponds to our Southern Gondwanian unit; (2) a Western North American unit, which corresponds to our

Siberian subunit; and (3) an Equatorial unit, which corresponds to our Equatorial and low-latitude subunit plus our Australian unit. Within their Equatorial unit, Ziegler et al. (1981) described three subunits: (1) an Appalachian subunit, which corresponds to our American subunit; (2) a Western European subunit, which corresponds approximately to our European subunit but which excludes assemblages from eastern Europe; and (3) an Eastern Equatorial subunit, which corresponds approximately to our Chinese and Kazakhstanian subunits and our Australian unit. In our study, by keeping floral assemblage lists separate during multivariate statistical analysis, we identified an additional phytogeographic unit (the Australian unit) and, more important, accurately evaluated the relationships among phytogeographic units and subunits.

Gedinnian–Early Siegenian Phytogeography

Although the number of assemblages of Gedinnian–early Siegenian age is too small for a multivariate statistical analysis, two phytogeographic units stand out. The assemblages from Australia described by Garratt (1978) share no genera with assemblages from more northern paleolatitudes. The age of these Australian assemblages is well constrained by marine-invertebrate biostratigraphy and lithostratigraphy. The disjunction between these Australian assemblages and other Gedinnian–early Siegenian assemblages suggests that floral assemblages from Australia belonged to a separate paleogeographic unit in the Gedinnian and possibly in the late Silurian as well (Garratt, 1978).

Without better sampling, we cannot evaluate the differentiation among the northern-hemisphere assemblages from this time interval. These occur between the north-temperate paleolatitude of 30°N (Spitsbergen) and the paleolatitude of 10°S (Scotland) on all three "northern-hemisphere" paleocontinents (Laurussia, Kazakhstan, and China).

Conclusions

Cluster analysis and polar ordination revealed three major phytogeographic units in the middle Siegenian and Emsian: (1) an equatorial and low-latitude phytogeographic unit; (2) an Australian phytogeographic unit; and (3) a Southern Gondwanian phytogeographic unit. Within the Equatorial and low-latitude phytogeographic unit, cluster analysis and polar ordination analysis revealed five intergrading subunits along a northwest-to-southeast paleogeographic gradient. These were: (1) the

Siberian subunit; (2) the American subunit; (3) the European subunit; (4) the Kazakhstanian subunit; and (5) the Chinese subunit.

During the Gedinnian and early Siegenian, regionally shared endemic genera suggest the presence of two phytogeographic units, an Australian unit and a wide-ranging "northern-hemisphere" unit.

Links between Early Devonian floral assemblages on the paleocontinent of Siberia and assemblages from equatorial paleolatitudes suggest that the paleocontinent of Siberia may have been further south in the Early Devonian than most paleogeographic reconstructions show.

Acknowledgments

This study owes its existence to the senior thesis of E. Gierlowski. The analytic techniques used were developed during discussions with J. J. Sepkoski, Jr., W. F. Schmachtenberg, and E. Gierlowski. A. M. Ziegler and C. R. Scotese provided paleogeographic reconstructions. A. Knoll, P. Gensel, A. Scott, and S. Schekler provided invaluable paleobotanical insight. A. Lottes provided bibliographic information and enthusiasm. M. D. Raymond and E. Gierlowski assisted in translating articles. J. Oliensis and B. Tiffney helped to edit this study; K. McDonald was tireless in her efforts to type it. Chevron Oil Company provided financial support.

Appendix

Table 6A.1 Floral Assemblages Used in Phytogeographic Analysis of the Middle Siegenian and Emsian

Location	Number on Figure 6.3	Abbreviation	Latitude	Longitude	Reference[*]
Sayan Altai, U.S.S.R.	1	NW-1	43N	96E	1
Minusinsk, U.S.S.R.	2	NW-2	53N	92E	1
Lena R., U.S.S.R.	3	NW-3	56N	93E	2
Wyoming, U.S.A.	14	NW-4	45N	110W	3,4
Spitsbergen, Norway	22	NW-5	79N	7E	5,6
Karasor, U.S.S.R.	4	K-1	50N	76E	7
Kazakhstan, U.S.S.R.	5	K-2	near 50N	76E	8
Germany-1	19	C-1	51N	7E	9
Germany-2	20	C-2	51N	8E	9
Long Hua, China	33	C-3	near 24N	110E	10
W. Yunnan, China	34	C-4	27N	101E	10
Wenshan, China	35	C-5	near 23N	104E	10
E. Kwangsi, China	36	C-6	near 24N	110E	10
E. Yunnan, China	37	C-7	25N	104E	10
Dushan, China	38	C-8	26N	108E	10
Duyun, China	39	C-9	26N	107E	10
Sahara	40	C-10	27N	7W	11
Molotov-1, U.S.S.R.	6	E-1	52N	46E	12
Molotov-2, U.S.S.R.	7	E-2	59N	57E	12
Podolia, U.S.S.R.	8	E-3	49N	26E	13
Dniester R., Ú.S.S.R.	9	E-4	49N	26E	14
Wales	16	E-5	52N	3W	15
Houyoux R., Belgium	18	E-6	50N	5E	16
Roragen, Norway	19	E-7	62N	6E	17
Meuse R., Belgium	26	E-8	52N	5E	16
Estinnes au Mont, Belgium	27	E-9	50N	4E	16
Carrière à Dave, Belgium	28	E-10	50N	5E	16
Bruay, Belgium	29	E-11	51N	5E	18,19
Matringham, Belgium	20	E-12	51N	2E	20
Andrychow, Poland	31	E-13	50N	19E	21,22
Barrancos, Portugal	32	E-14	33N	8W	23,24
New Brunswick, Canada	10	A-1	48N	66W	25
Gaspé Bay, Canada	11	A-2	48N	66W	25

Table 6A.1   (Continued)

Location	Number on Figure 6.3	Abbreviation	Latitude	Longitude	Reference*
Maine, U.S.A.	12	A-3	46N	69W	25
James Bay, Canada	13	A-4	51N	81W	26,27
Perth, Scotland	17	A-5	56N	4W	28,29
Cardross, Scotland	24	A-6	56N	5W	30
Oxford, England	25	A-7	52N	1W	31
Port Alfred, South Africa	41	G-1	33S	27E	32
Bathurst District, South Africa	42	G-2	33S	27E	32
Ross Sea, Antarctica	46	G-3	73S	106E	32
Mt. Pleasant, Australia	43	Aus-1	37S	144E	33
Yarra Track, Australia	44	Aus-2			34
Lilydale, Australia	45	Aus-3	38S	145E	35
Bathurst Island, Canada	15	?-1	77N	99W	4
Harz Mt., East Germany	21	?-2	52N	11E	37

* Reference numbers refer to the following articles: (1) Ananiev and Krasnov, 1962; (2) Krylova et al., 1967; (3) Dorf, 1934; (4) Hueber, 1971a; (5) Høeg, 1942; (6) Schweitzer, 1968; (7) Chetverikova et al., 1971; (8) Petrosyan, 1967; (9) Kräusel and Weyland, 1930, 1932, 1935; (10) Li and Cai, 1978; (11) Lemoigne, 1967; (12) Tschirkova-Zalesskaja, 1954; (13) Ischenko, 1965; (14) Ischenko, 1974; (15) Croft and Lang, 1942; (16) Stockmans, 1940; (17) Halle, 1917; (18) Corsin, 1933; (19) Danzé-Corsin, 1956; (20) Danzé-Corsin, 1955; (21) Jakubowska, 1968; (22) Zdebska, 1972; (23) Teixeria, 1951; (24) Perdigao, 1967; (25) Andrews et al., 1977; (26) Lemon, 1953; (27) Hueber, 1964; (28) Henderson, 1932; (29) Lang, 1932; (30) Scott, Edwards, and Rolfe, 1976; (31) Chaloner and Sheerin, 1979; (32) Plumstead, 1967; (33) Gould, 1975; (34) Cookson, 1945; (35) Cookson, 1949; (36) Douglas and Ferguson, 1976; (37) Magdefrau, 1938.

Table 6A.2 · Genera Counted as Present in Middle Siegenian–Emsian Assemblages (1 = Presence)

Assemblage	NW-1	NW-2	NW-3	NW-4	NW-5	K-1	K-2	C-1	C-2	C-3	C-4	C-5	C-6	C-7	C-8	C-9	C-10
Baragwanathia	1	1	1	1	1												
Bucheria/Rebuchia																	1
Cooksonia				1	1	1											
Drepanophycus			1					1	1	1		1		1	1	1	
Eogaspesia														1		1	
Gosslingia																	
Haplostigma							1										
Hedia																	
Kaulangiophyton							1										
Liadesmophyton						1	1										
Paleostigma																	
Pertica																	
Protobarinophyton/Barinophyton	1	1	1		1		1										
Protolepidodendron	1	1					1	1									
Psilophyites						1				1					1		
Psilophyton	1	1	1	1	1	1					1						
Sawdonia				1													
Sciadophyton								1									
Sporogonites						1		1									
Taeniocrada	1	1				1	1	1	1	1	1	1	1	1	1	1	1
Tomiophyton	1	1				1											
Yarravia																	
Zosterophyllum					1	1	1	1	1	1	1	1	1	1	1	1	1

Table 6A.2 (Continued)

Assemblage	E-1	E-2	E-3	E-4	E-5	E-6	E-7	E-8	E-9	E-10	E-11	E-12	E-13	E-14	A-1	A-2	A-3	A-4
Baragwanathia	1																	
Bucheria/Rebuchia																		
Cooksonia					1													
Drepanophycus				1	1	1	1	1	1	1	1	1	1	1	1	1		1
Eogaspesia		1	1													1		
Gosslingia		1	1	1														
Haplostigma																		
Hedia																		
Kaulangiophyton														1				
Liadesmophyton																	1	
Paleostigma																		
Pertica																1	1	
Protobarinophyton/Barinophyton													1					
Protolepidodendron								1										
Psilophytites																		
Psilophyton	1	1	1	1	1	1	1	1	1	1	1	1	1	1	1	1	1	1
Sawdonia			1	1	1			1		1					1	1	1	1
Sciadophyton			1	1	1													
Sporogonites			1	1	1	1		1	1	1								
Taeniocrada	1	1	1	1	1		1	1	1	1	1		1	1	1		1	
Tomiophyton		1	1															
Yarravia										1								
Zosterophyllum		1	1		1	1				1					1			

Table 6A.2 (Continued)

Assemblage	A-5	A-6	A-7	G-1	G-2	G-3	Aus-1	Aus-2	Aus-3	?-1	?-2
Baragwanathia							1				
Bucheria/Rebuchia		1						1		1	1
Cooksonia	1										
Drepanophycus				1						1	1
Eogaspesia				1	1						
Gosslingia											
Haplostigma				1	1						
Hedia								1	1		
Kaulangiophyton											
Liadesmophyton											
Paleostigma				1	1						
Pertica				1							
Protobarinophyton/Barinophyton						1					
Protolepidodendron											
Psilophyites											
Psilophyton		1	1								
Sawdonia	1	1	1								
Sciadophyton									1	1	
Sporogonites											
Taeniocrada											
Tomiophyton											
Yarravia							1	1	1		
Zosterophyllum							1	1	1		

References

Ananiev, A. R., and E. Krasnov. 1962. Stratigraphy of the Devonian in the Tustuchzhul syncline, south Minusinsk basin [in Russian]. *Dokl. Akad. Nauk CCCP* 145:867–70.

Andrews, H. N., Jr. 1961. Some Paleozoic and Mesozoic floras. In *Studies in fossil botany*, ed. H. N. Andrews, Jr., 408–12. New York: John Wiley and Sons.

Andrews, H. N., A. E. Kasper, W. H. Forbes, P. G. Gensel, and W. G. Chaloner. 1977. Early Devonian flora of the Trout Valley Formation of northern Maine. *Rev. Palaeobot. Palynol.* 23:255–85.

Arber, E. A. N. 1921. *Devonian floras.* Cambridge: Cambridge Univ. Press.

Bailey, I. W., and E. W. Sinnott. 1915. A botanical index of Cretaceous and Tertiary climates. *Science* 41:831–34.

———. 1916. The climatic distribution of certain types of angiosperm leaves. *Am. J. Bot.* 3:24–39.

Banks, H. P. 1980. Floral assemblages in the Siluro-Devonian. In *Biostratigraphy of fossil plants*, ed. D. L. Dilcher and T. N. Taylor, 1–24. Stroudsburg, PA: Dowden, Hutchinson and Ross.

Beals, E. W. 1965. Ordination of some corticolous cryptogamic communities in south-central Wisconsin. *Oikos* 16:1–8.

Bray, J. R., and J. T. Curtis. 1957. An ordination of the upland forest communities of southern Wisconsin. *Ecol. Monogr.* 27:325–49.

Chaloner, W. G., and W. S. Lacey. 1973. The distribution of Late Paleozoic floras. In *Organisms and continents through time*, ed. N. F. Hughes, 271–89. Special Papers in Palaentology, 12. London: Palaeontological Association.

Chaloner, W. G., and S. V. Meyen. 1973. Carboniferous and Permian floras of the northern continents. In *Atlas of paleobiogeography*, ed. A. Hallam, 169–86. Amsterdam: Elsevier Scientific Publishing Co.

Chaloner, W. G., and A. Sheerin. 1979. Devonian macrofloras. In *The Devonian system*, ed. M. House, 145–61. Special Papers in Palaeontology, 23. London: Palaeontological Association.

Cheetham, A. H., and J. E. Hazel. 1969. Binary (presence-absence) similarity coefficients. *J. Paleontol.* 43:1130–36.

Chetverikova, N. P., O. E. Belyaev, Yu. F. Kabanov., G. T. Ushatinskaya, and B. Ya. Zhuravlev. 1971. Stratigraphy of the Devonian deposits in the northern parts of the Hercynides of the central Kazakhstan and marginal volcanic belt [in Russian]. *Vestn. Mosk. Univ., Ser. 4, Geol.* 1:3–16.

Cisne, J. L., and B. D. Rabe. 1978. Coenocorrelation: Gradient analysis of fossil communities and its application in stratigraphy. *Lethaia* 11:341–64.

Cookson, I. C. 1935. On plant remains from the Silurian of Victoria, Australia, that extend and connect floras hitherto described. *Philos. Trans. R. Soc. London* 225B:127–48.

———. 1945. Records of plant remains from the Upper Silurian and Early Devonian rocks of Victoria. *Proc. R. Soc. Victoria* 56:119–22.

————. 1949. Yeringian (Lower Devonian) plant remains from Lilydale, Victoria, with notes on a collection from a new locality in the Siluro-Devonian sequence. *Mem. Natl. Mus. Victoria* 16:117–31.

Corsin, P. 1933. Découverte d'une flore dans le Dévonien inférieur de Pas-de-Calais. *C. R., Aca. Sci. Inst. France* 197:180–181.

Croft, W. N., and W. H. Lang. 1942. The Lower Devonian floras of the Senni Beds of Monmouthshire and Breconshire. *Philos. Trans. R. Soc. London* 231B:131–63.

Danzé-Corsin, P. 1955. Contribution a l'étude des flores dévoniennes du Nord de la France. I. — Flore éodévonienne de Matringhem. *Ann. Soc. Géol. Nord* 75:143–60.

————. 1956. Contribution a l'étude des flores dévoniennes du Nord de la France. II. — Flore éodévonienne de Rebreuve. *Ann. Soc. Géol. Nord* 76:24–50.

Dorf, E. 1934. Stratigraphy and paleontology of a new Devonian formation at Beartooth Butte, Wyoming. *J. Geol.* 42:720–37.

Dorf, E., and D. W. Rankin. 1962. Early Devonian plants from the Traveler Mountain area, Maine. *J. Paleontol.* 36:999–1004.

Douglas, J. G., and J. A. Ferguson. 1976. The geology of Victoria. *Spec. Publ. — Geol. Soc. Aust.* 5:56.

Edwards, D. 1970. Fertile Rhyniophytina from the Lower Devonian of Britain. *Palaeontology* 13:451–61.

————. 1972. A *Zosterophyllum* fructification from the Lower Old Red Sandstone of Scotland. *Rev. Palaeobat. Palynol.* 13:77–83.

————. 1973. Devonian floras. In *Atlas of palaeobiogeography*, ed. A. Hallam, 105–15. New York: Elsevier Scientific Publishing Co.

Edwards, D., and J. B. Richardson. 1974. Lower Devonian (Dittonian) plants from the Welsh borderland. *Paleontology* 17:311–24.

Garratt, M. J. 1978. New evidence for a Silurian (Ludlow) age for the earliest *Baragwanathia* flora. *Alcheringa* 2:217–24.

Gierlowski, E. H. 1978. On the geographic distribution of vascular plants in the Devonian. B. S. thesis, Univ. of Chicago.

Gould, R. E. 1975. The succession of Australian pre-Tertiary megafossil floras. *Bot. Rev.* 41:453–83.

Halle, T. B. 1917. Lower Devonian plants from Roragen in Norway. *Kungliga Svenska Vetenskapsak Academiens Handlingar* 57:1–46.

Henderson, S. M. K. 1932. Notes on the Lower Old Red Sandstone plants from Callander, Perthshire. *Trans. R. Soc. Edinburgh* 57:277–86.

Høeg, O. A. 1942. *The Downtonian and Devonian flora of Spitzbergen.* Norges Svalbard-og Ishava-Undersøkelser Skrifter Nr. 83. I Kommisjon Hos Jacob Dybwad. Oslo.

Hueber, F. M. 1964. The psilophytes and their relationship to the origin of ferns. *Mem. Torrey Bot. Club* 21:5–8.

————. 1967. *Psilophyton:* The genus and the concept. In *International symposium on the Devonian system*, ed. D. H. Oswald, 2:815–22. Calgary: Alberta Society of Petroleum Geologists.

———. 1971a. Early Devonian land plants from Bathurst Island, District of Franklin. *Geol. Surv. Pap. (Geol. Surv. Can.)*, no. 71–28.

———. 1971b. *Sawdonia ornata:* A new name for *Psilophyton princeps var. ornatum. Taxon* 20:641–42.

Ischenko, T. A. 1965. Devonian flora of the Volynian-Podolian margin of the Russian platform [in Russian]. *Paleontologischeskii Sbornik* 2:123–25.

———. 1974. *Tirasophyton*, a new Early Devonian plant genus from Podolia [in Russian]. *Paleontologischeskii Zhurnal* 1:112–14.

Jakubowska, L. 1968. Paleobotanical stratigraphic studies on Devonian deposits pierced by boreholes, Ciepielów and Dorohucza [in Polish]. *Kwartalnik Geologiczny* 12:507–18.

Kräusel, R., and H. Weyland. 1930. Die flora des deutschen Unterdevons. *Preussischen Geologischen Landesanstalt* 131:1–90.

———. 1932. Pflanzenreste aus dem Devon, V. Zwei unterdevonische Pflanzenrhizome. *Senckenbergiana* 14:403–06.

———. 1935. Neue Pflanzenfunde im Rhenischen Unterdevon. *Palaeontographica* 80B:171–90.

Krassilov, V. A. 1975. *Paleoecology of terrestrial plants.* Trans. H. Hardin. New York: John Wiley and Sons.

Krylova, A. K., N. S. Maltich, V. V. Menner, D. V. Obrutchev, and S. G. Fradkin. 1967. Devonian of the Siberian Platform. In *International symposium on the Devonian system*, ed. D. H. Oswald, 1:473–82. Calgary: Alberta Society of Petroleum Geologists.

Lang, W. H. 1932. Contributions to the study of the Old Red Sandstone flora of Scotland. VIII. On *Arthrostigma, Psilophyton*, and some associated plant-remains from the Strathmore Beds of the Caledonian Lower Old Red Sandstone. *Trans. — R. Soc. Edinburgh* 57:491–521.

Leclercq, S. 1942. Quelques plantes fossiles recueillies dans le Dévonien inférieur des environs de Nonceveaux (Bordure orientale du bassin de Dinant). *Ann. Soc. Géol. Belg.* 65:193–211.

Lemoigne, Y. 1967. Reconnaissance paléobotanique dans Le Sahara occidental (Région de Tinouf et Gara-Djebilet). *Ann. Soc. Géol. Nord* 87:31–38.

Lemon, R. R. H. 1953. The Sextant Formation and its flora. Master's thesis, Univ. of Toronto.

Li Zing-zue and Cai Chong-yang. 1978. A type-section of Lower Devonian strata in southwest China with brief notes on the succession and correlations of its plant assemblages [in Chinese]. *Acta Geol. Sin.* 1:1–14.

MacArthur, R. H., and E. O. Wilson. 1967. *The theory of island biogeography.* Monographs in Population Biology, 1. Princeton, NJ: Princeton Univ. Press.

Magdefrau, K. 1938. Eine Halophyten-Flora aus dem Unterdevon des Harzes. *Beihefte zum Botanischen Centralblatt* 58B:243–51.

Orloci, L. 1978. *Multivariate analysis in vegetation research.* The Hague: Junk.

Perdigao, J. C. 1967. Sobre o prolongamento e presumivel idade da faixa com vegetais fosseis de Eiras Altas (Barrancos). *Comun. Serv. Geol. Port.* 52:49–54.

Petrosyan, N. M. 1967. Stratigraphic importance of the Devonian floras of the U.S.S.R. In *International symposium on the Devonian system*, ed. D. H. Oswald, 2:579–86. Calgary: Alberta Society of Petroleum Geologists.

Pianka, E. R. 1974. *Evolutionary ecology*. New York: Harper and Row.

Plumstead, E. P. 1967. A general review of the Devonian fossil plants found in the Cape System of South Africa. *Palaeontologia Africana* 10:1–83.

Schimper, A. F. W. 1903. *Plant geography on a physiological basis*. Trans. W. R. Fisher; ed. P. Groom and I. B. Balfour. Oxford: Oxford Univ. Press.

Schmachtenberg, W. F., Jr. 1983. The geological implications of a study of an Upper Cretaceous epicontinental seaway fauna. Ph.D. diss., Univ. of Chicago.

Schmidt, W. 1954. Pflanzenreste aus der Siegerlandes. I. *Sugambrophyton pilgeri* n.g., n.s., eine Protolepidodrendracea aus den Hamburg-Schichten. *Palaeontographica* 97B:1–22.

Schweitzer, H. J. 1968. Pflanzenreste aus dem Devon, nord-west Spitzbergen. *Palaeontographica* 123B:43–75.

Scotese, C. R., R. K. Bambach, C. Barton, R. Van der Voo, and A. M. Ziegler. 1979. Paleozoic base maps. *J. Geol.* 87:217–77.

Scott, A. C., D. Edwards, and W. D. I. Rolfe. 1976. Fossiliferous Lower Old Red Sandstone near Cardross, Dunbartonshire. *Proc. Geol. Soc. Glasgow* 117:4–5.

Smith, A. G., A. M. Hurley, and J. C. Briden. 1981. *Phanerozoic paleocontinental world maps*. Cambridge Earth Science Series. Cambridge: Cambridge Univ. Press.

Sneath, P. M. A. and R. R. Sokal. 1973. *Numerical taxonomy*. San Francisco: W. H. Freeman.

Spicer, R. A. 1980. The importance of depositional sorting to the biostratigraphy of plant megafossils. In *Biostratigraphy of fossil plants*, ed. D. L. Dilcher and T. N. Taylor, 171–84. Stroudsburg, PA: Downden, Hutchinson and Ross.

Stockmans, F. 1940. Végetaux éodévoniens de la Belgique. *Mem. R. Hist. Nat. Belg.* 93:1–90.

Teixeria, C. 1951. Notas sobre a geologica da regiao de Barrancos e, em especial, sobre a sua flora de Psilofitnineas. *Comun. Serv. Geol. Port.* 32:75–83.

Tschirkova-Zalesskaja. E. G. 1954. Iskopaem'ie rasteniya terrigennogo Devona Uralo-Povolzh'ya. *Dokl. Akad. Nauk CCCP* 94:129–32.

Walter, H. 1973. *Vegetation of the earth in relationship to climate and the ecophysiological conditions*. Trans. J. Wieser. New York: Springer-Verlag.

Whittaker, R. H., ed. 1978. *Ordination of plant communites*. The Hague: Junk.

Zdebska, D. 1972. *Sawdonia ornata* (*Psilophyton princeps var. ornatum*) from Poland. *Acta Palaeobotanica* 13:77–99.

Ziegler, A. M., R. K. Bamhach, J. T. Parrish, S. F. Barrett, E. H. Gierlowski, W. C. Parker, A. Raymond, and J. J. Sepkoski, Jr. 1981. Paleozoic biogeography and climatology. In *Paleobotany, paleoecology, and evolution*, ed. K. J. Niklas, 2:231–66. New York: Praeger.

ANNE RAYMOND, WILLIAM C. PARKER, AND
JUDITH TOTMAN PARRISH

Phytogeography and Paleoclimate of the Early Carboniferous

Introduction

In this chapter, we discuss the compression floral phytogeography and paleoclimate of two Early Carboniferous time intervals, Tournaisian to early Visean and late Visean to earliest Namurian A. For convenience, we will use the term *differentiation* to refer to the biogeographic differentiation of compression floral, palynofloral, and marine-invertebrate assemblages. Our study differs from previous studies in that we use detailed paleogeographic reconstructions and multivariate statistical techniques to establish phytogeographic units. Previous studies relied on the identification of regionally endemic taxa to define phytogeographic units (Jongmans, 1954; Chaloner and Meyen, 1973; Chaloner and Lacey, 1973). These studies suggested that very little differentiation existed during the Early Carboniferous; however, if we assume that Hadley circulation and climatic variation have been constant throughout the Phanerozoic, this conclusion appears counterintuitive. Because Early Carboniferous landmasses span such a wide area (from paleolatitude 80°N to 80°S), we expected Early Carboniferous assemblages to show some differentiation.

Using the multivariate statistical technique of polar ordination, we find evidence for five Tournaisian–early Visean and three late Visean –earliest Namurian A phytogeographic units. We evaluate these results using paleoclimatic models presented in this paper. These models explain the differentiation observed in Tournaisian–early Visean

169

assemblages fairly well. However, for the late Visean and earliest Namurian A, our model predicts a degree of paleoclimatic differentiation that we do not observe in the results of our phytogeographic study.

Our results are consistent with those of other studies of Early Carboniferous palynofloral phytogeography and marine-invertebrate biogeography. As a whole, the results of these studies suggest that when the seaway between Laurussia and Gondwana disappeared, the amount of floral differentiation decreased due to climatic amelioration along the east coast of Pangaea (Laurussia plus Gondwana).

This investigation is one in a series on the phytogeography of the Paleozoic and Mesozoic compression floral assemblages. It differs from previous studies of the Early Carboniferous in the use of narrower time intervals, more detailed paleogeographic reconstructions, and more powerful analytical techniques.

Previous Work

Jongmans (1952, 1954), Rigby (1969), Vakhrameev et al. (1970), Chaloner and Meyen (1973), Chaloner and Lacey (1973), and Ziegler et al. (1981) have published phytogeographic analyses of Early Carboniferous compression floral assemblages. Jongmans (1952, 1954) believed that all Early Carboniferous assemblages belonged to the same phytogeographic province on the basis of the worldwide distribution of the genera *Lepidodendropsis*, *Rhacopteris*, and *Triphyllopteris*; however, he worked before the acceptance of continental drift and without the benefit of detailed biostratigraphy. His floral assemblages span a long time interval, from the Dinantian (Tournaisian and Visean) through the early Namurian (Jongmans, 1952).

Rigby (1969) reevaluated the Early Carboniferous compression floral assemblages of the Gondwana continents (South America, Africa, India, and Australia). He concluded that the *Rhacopteris* assemblages of these continents constituted a phytogeographic province of Visean or later age, which was separate from the northern-hemisphere flora of the same interval. Archangelsky (1983) later assigned southern-hemisphere *Rhacopteris* species to the genus *Nothorhacopteris*.

Vakhrameev et al. (1970) described two phytogeographic provinces in the Early Carboniferous of Eurasia: a European province consisting of assemblages from western Europe, England, the Urals, Kazakhstan, and China; and an Angaran province consisting of assemblages from Siberia and Mongolia.

Chaloner and Meyen (1973) reevaluated the distribution of Tournaisian, Visean, and early Namurian compression genera and suggested the existence of two or possibly three phytogeographic units during this time interval, based on the occurrence of regionally endemic genera. The first of these units included assemblages from Africa, Australia, and South America (the paleocontinent of Gondwana) and from the United States, the Canadian Maritimes, Europe, and China (the paleocontinents of Laurussia and China). Chaloner and Meyen (1973) felt that the apparent similarity between Laurussian and Gondwanian assemblages might be due to the application of "broad" generic concepts by workers in the field. The second unit included floral assemblages from Siberia and Mongolia. The authors also recognized a possible third unit composed of Central Asian assemblages from the paleocontinent of Kazakhstan.

Chaloner and Lacey (1973) described the same two or possibly three Early Carboniferous (Tournaisian through Namurian) phytogeographic units: the *Lepidodendropsis* flora found on the paleocontinents of Laurussia, Gondwana, and China; the Angaran flora found in present-day Siberia and Mongolia; and a possible Kazakhstanian flora composed of Central Asian floral assemblages. They mapped these units on a paleogeographic reconstruction provided by Smith, Briden, and Drewry (1973). In this reconstruction, the Acadian terrains (Great Britain and a portion of the Canadian Maritimes) were shown in their present-day positions relative to North America and Europe. In addition, this reconstruction shows Siberia, Kazakhstan, and China joined into one paleocontinent.

The studies by Jongmans (1952, 1954), Chaloner and Meyen (1973), and Chaloner and Lacey (1973) span three stages (Tournaisian, Visean, and early Namurian) and were produced without reference to detailed Early Carboniferous paleogeographic reconstructions. These workers relied on the identification of regionally shared endemic taxa, such as *Lepidodendropsis* for Laurussia, Gondwana, and China, *Chacassopteris* and *Angaropteridium* for Siberia, and *Caenodendron* for Kazakhstan, in order to delineate phytogeographic provinces.

Ziegler et al. (1981) produced a preliminary study of Early Carboniferous phytogeography using the techniques outlined in this study. They attempted to limit their analysis to Visean floral assemblages and used detailed paleogeographic reconstructions. They found evidence of four Early Carboniferous phytogeographic units: an Angaran unit (consisting of assemblages from the paleocontinent of Siberia); a Gondwanian unit; a Northern Circum-Tethys unit; and a Southern Circum-Tethys unit. The Angaran unit included Mongolian assemblages and contained only late Visean and earliest Namurian A floral assemblages. However, the Northern Circum-Tethys unit, the Southern

Circum-Tethys unit, and the Gondwanian unit included assemblages that ranged from Strunian through early Namurian A in age. Thus the possibility remains that dissimilar ages as well as phytogeographic differentiation formed the basis for the Early Carboniferous phytogeographic units described their study.

Data Analysis

The paleophytogeographic analyses presented here required three steps. First, we had to identify differentiation among floral assemblages. We did this using polar ordination, a multivariate statistical technique that groups assemblages based on the number of shared taxa and that has proved a powerful tool for paleobiogeographic analysis of marine-invertebrate distributions (Crick, 1980; Whittington and Hughes, 1972). We discuss the aspects of Early Carboniferous paleobotany, paleogeography, and paleoclimate that complicate phytogeographic analysis and the ways in which ordination analysis circumvents these difficulties in the section on Early Carboniferous phytogeographic analysis. We outline the procedures involved in polar ordination in the section on statistical methods.

Second, we had to separate the differentiation caused by paleogeography from that caused by the comparison of floral asemblages with dissimilar ages and habitats. To accomplish this, we tried to gather a uniform data set that eliminated as many sources of nonphytogeographic variation as possible. We discuss some of these sources of variation in the sections on standards for the inclusion of assemblages and standards for the inclusion of taxa.

Third, we had to analyze the remaining differentiation to see if it could be explained by paleogeography. We used detailed paleogeographic reconstructions produced by A. M. Ziegler and C. R. Scotese at the University of Chicago and studies of Early Carboniferous marine biogeography to interpret our results. We also tested our results against paleoclimatic models presented below. The interpretation of paleophytogeographic results requires both paleogeographic reconstructions and paleoclimatic models because of the convergent evolution of plants in response to climate. We discuss this problem in the following section.

Interpretation

The geographic distribution of plant taxa forms the data base for both modern and ancient phytogeography. However, the interpretation of ancient phytogeography differs from that of modern phytogeography because ancient and modern plant taxa are not always comparable. To the extent that modern plant taxonomy, based on reproductive characters, reflects phylogeny, phylogenetic differences determine modern phytogeographic units. The majority of ancient plant compression taxa are defined using vegetative morphology, and reproductive structures assume systematic importance only when available. Vegetative characters are particularly subject to convergent evolution (Chaloner and Meyen, 1973). Because vegetative morphology as well as reproductive characters are used to determine ancient taxa, these may contain biologically unrelated forms. Thus, phytogeographic units based on ancient taxa may be determined by non-phylogenetic convergence and phylogenetic factors.

We assume that morphological separation implies the presence of some genetic separation between phytogeographic units. Therefore we believe that ancient phytogeographic units reflect biogeographic differentiation in the modern sense. However, the units delineated represent a conservative view of the amount of differentiation present. Each ancient unit may include more than one "phylogenetic" biogeographic unit.

Climate influences the morphology of modern plants, as indicated by the field of eco-physiology founded by Schimper (1903) and continued by workers such as Bailey and Sinnot (1915) and Walter (1973). If any relationship between climate and plant morphology obtained in the Paleozoic, then ancient phytogeographic units reflect a mixture of paleoclimatic and phylogenetic factors. Ancient and modern plants almost certainly had different morphological adaptations to climate. In particular, individuals of some angiosperm species possess different leaf shapes in different habitats. The morphological adaptations of Paleozoic plants probably represent species-level changes rather than variability within a species. We assume only that ancient plants responded to climatic stress by modifying their vegetative morphology; we make no assumption as to the form of their response. Given this assumption, paleoclimatic models are essential for the interpretation of ancient phytogeography.

Figure 6.1 in Raymond, Parker, and Barrett (this volume) illustrates one possible relationship between modern and ancient phytogeographic units. In it, two hypothetical modern phytogeographic units (separated by a vertical line) contain three subunits corresponding to different climatic zones. Three hundred fifty million years in the future, a phytogeographic analysis of the same region might yield three ancient phytogeographic units (separated by horizontal lines) corresponding to the three original climatic zones.

The relationship of individual assemblages within modern and ancient phytogeographic units also differs. Modern phytogeographic units are hierarchical. All the assemblages within subunit A_1 are more closely genetically related to one another than to any assemblage in subunit A_2. All the assemblages in modern unit A are more closely genetically related to one another than to any assemblage in modern unit B. Ancient phytogeographic units are not necessarily hierarchical. The assemblages that make up ancient unit 1 are not more closely related to one another than to the assemblages of ancient units 2 and 3. Morphological convergence in response to climate may lead to the interpretation that two morphologically similar assemblages such as A_1 and B_1 formed a biogeographic unit (ancient unit III) when they actually have different genetic heritages.

Early Carboniferous Phytogeographic Analysis

Certain aspects of Early Carboniferous paleobotany and paleogeography may obscure differentiation during this interval. Among these aspects are: (1) the low generic diversity of Early Carboniferous floral assemblages; (2) the presence of gentle climatic gradients during the Early Carboniferous; and (3) poorly understood paleogeography. The Early Carboniferous shares many of these features with other periods, particularly the Devonian, and the following discussion also bears on the phytogeographic analysis of Devonian floras. Our approach to phytogeographic analysis may circumvent some of these problems.

Low Generic Diversity. During times of lower generic diversity, it is possible that individual plant genera may have had broader climatic ranges. Thus there may have been few endemic taxa that distinguished phytogeographic units. Latitudinal and longitudinal gradients in the distribution of plant genera during such periods still might reveal differentiation. However, if gradational distributions alone distinguish phytogeographic units, techniques that rely on the identification of regional endemics will not uncover this differentiation. Ordination analysis, which groups assemblages by overall floral similarity, can uncover phytogeographic gradients among floral assemblages with low generic diversity.

Gentle Latitudinal Temperature Gradients. Traditional techniques of phytogeographic analysis work best when comparing paleoclimatically disjunct regions. Siberia, one of the only Early Carboniferous phytogeographic provinces recognized using traditional techniques of analysis, lay in cold temperate latitudes, paleoclimatically disjunct from all other Early Carboniferous land areas except southern Gondwana.

Due to the presence of open polar oceans, the Early Carboniferous may have had a gentle latitudinal climatic gradient (Robinson, 1973). During such times, there may be few endemic taxa that distinguish phytogeographic units. However, as with times of low generic diversity, latitudinal or longitudinal gradients in the distribution of taxa may still reflect differentiation. Ordination analysis may uncover this gradational differentiation.

Poorly Understood Paleogeography. In the Early Carboniferous, the geographic position of many regions is inexactly known. Among these regions are the terrains involved in the Acadian orogeny (portions of the Canadian Maritime provinces and Scotland, England, and Ireland) and parts of Europe south of the Hercynian orogenic belt (Scotese et al., 1984). Obviously, lumping floral assemblages from these poorly placed regions with assemblages from the nearby stable cratons could obscure paleophytogeographic differences. Alternatively, if paleophytogeographic differences are noted, they may be ignored as meaningless in the absence of accurate information about the positions of the poorly placed terrains. For example, during the Early Ordovician North Africa, Spain, and Florida share a trilobite fauna characterized by *Selenopeltis*, while the rest of North America contains a bathyurid fauna (Whittington and Hughes, 1972). Ordovician paleogeographic reconstructions that show Florida between Africa and South America (Ziegler et al., 1981) aid in the interpretation of Early Ordovician trilobite biogeography. Lumping assemblages from Florida with those from the rest of North America would obscure biogeographic differentiation. In order to avoid the problems posed by poorly understood paleogeography, we used detailed paleogeographic reconstructions. We also used studies of Early Carboniferous marine biogeography. These clarify the paleogeography of poorly understood regions by indicating the presence of land barriers.

Conclusions. To summarize, multivariate statistical techniques, such as polar ordination analysis, can circumvent the problems posed by phytogeographic analysis of time periods with low generic diversity and gentle latitudinal gradients. Poorly understood paleogeography poses the most difficult problem in phytogeographic analyses. These analyses require detailed paleogeographic reconstructions and an awareness of poorly understood regions. During the Early Carboniferous, marine biogeography provides additional evidence for the presence of Early Carboniferous land barriers.

Standards for the Inclusion of Assemblages

Lists of floral assemblages used in this study appear in the Appendix. In the following two sections we discuss our criteria for including floral assemblages and individual taxa in this study. We selected these criteria to eliminate as many sources of nonphytogeographic variation as possible.

For this study, we originally intended to collect only those floral lists reporting assemblages of fossil plants from a single bedding plane in a specific locality. In practice, whole paleocontinents would have been excluded from the analysis had we applied these standards rigorously. In order to include assemblages from the U.S.S.R. and Mongolia (the paleocontinents of Kazakhstan and Siberia) and China (the paleocontinents of North China and South China), we accepted assemblage lists from single formations within a limited geographic area. Including assemblages such as these introduces three possible sources of error: (1) consideration of plants from different habitats as if they formed a single assemblage with one habitat; (2) consideration of plants from widely separated stratigraphic horizons as a single assemblage; and (3) consideration of plants that were originally borne on separate microplates as members of a single assemblage. Of the three possible errors, the first two seem most likely.

Many factors besides paleogeography and paleoclimate can cause floral assemblages to appear different. Among these are differences in mode of preservation, geological age, habitat, and sampling intensity. In gathering the data for this study, we attempted to eliminate as many sources of nonphytogeographic differentiation as possible. We discuss these sources of differentiation below.

Mode of Preservation. Compression assemblages and permineralized assemblages representing the same community generally contain different organs, each with its own generic name. Even when these two types of assemblages contain the same organ from the same type of plant, the organ may have different generic names in each assemblage due to the difference in preservation. In order to prevent differentiation of floral assemblages due to mode of preservation, we followed the convention of Chaloner and Meyen (1973) and Chaloner and Lacey (1973) and limited our study to compression assemblages.

Assemblages of Different Ages. Lumping assemblages of different ages into the same phytogeographic analysis can either create false differentiation or obscure existing differentiation. Consideration of a large time interval may cause assemblages to be grouped together solely because they share the same time interval. Because assemblages of the same interval often come from the same region, such differentiation can mimic phytogeographic differentiation. In the analysis of Early Carboniferous phytogeography presented by Ziegler et al. (1981), assemblages from southern Europe and Virginia contained taxa that range from Strunian through Namurian in age (A. Scott, pers. com., 1981); the other assemblages used were Visean or earliest Namurian in age. In the analysis, the European assemblages with Strunian taxa grouped together; only the inclusion of an assemblage from Virginia distinguished this age-related differentiation of assemblages from phytogeographic differentiation.

Phytogeographic analyses of short time intervals may reveal differences obscured in analyses done for longer time intervals. This happens because plant taxa can migrate considerable distances over millions of years in response to changing climates or biological interactions. If the distribution of a paleogeographically restricted taxon changes during the time interval under consideration, floral assemblages of different ages from different climatic zones may appear to share a taxon that was regionally endemic to each climatic zone at separate times. The Mesozoic plant *Nilsonnia* is an example of a taxon that has changed its geographic distribution in geologic time by vacating its original climatic zone and invading another (Vakhrameev et al., 1970). *Nothorhacopteris-Rhacopteris* may represent a leaf shape that was largely tropical in its Tournaisian distribution and largely temperate in its post-Visean distribution (Rigby, 1969). For this reason, paleophytogeographic analyses for the Tournaisian and for the Namurian A should yield results different from those of an analysis of the entire Early Carboniferous.

Plant taxa can also vacate one climatic zone and invade another within hundreds or thousands of years. However, these geologically instantaneous migrations are beyond the resolution of the Paleozoic fossil record. In a paleophytogeographic analysis, such taxa will always appear to occur in both climatic zones.

Finally, if a continental plate shifted paleolatitude during the time interval under consideration, floral assemblages that grew in different climatic zones at different times may become lumped together in the analysis, obscuring climatically controlled floral differentiation. The use of the narrowest possible time intervals in paleophytogeographic analyses helps control these problems.

In order to eliminate differentiation of floral assemblages due to age, we divided our Early Carboniferous data into two time intervals, the Tournaisian–early Visean and the late Visean–earliest Namurian A. These were chosen to include some assemblages from the paleoconti-nents of Gondwana and Siberia (present-day Siberia and Mongolia) in each sample. We attempted to limit our earliest Carboniferous analysis to Tournaisian floral assemblages. However, the following assemblages could range into the earliest Visean: (1) the Jurong assemblage (Yang et al., 1979); (2) the Uralian assemblage, Bredy (Tchirkova, 1937; A. Scott, pers. com., 1981); (3) the Moroccan assemblages, Djebel Bakach and El Magnounan (Danzé-Corsin, 1960; A. Scott, pers. com., 1981); (4) the Kashmir assemblage (Pal and Chaloner, 1979); (5) the Mongo-lian assemblages (Durante, 1976).

Whenever possible, we confirmed the age of the floral assemblages used with either marine invertebrates or palynofloras. We excluded Early Carboniferous floral assemblages from West Virginia and Penn-sylvania (the Pocono and Price Formations) described by White (1913), Read (1955), Gensel (1979), Gensel and Skog (1977), Scheckler (1979) and Jennings (1979) and assemblages from basins of Hercynian age in southern Europe described by Bureau (1914) and Hirmer (1940). Published descriptions of assemblages from both areas may contain plants from more than one Lower Carboniferous stage (Strunian through Visean) (A. Scott, pers. com., 1981). Further, no apparent consensus exists as to the compression taxa present in either location, which further complicates the task of identifying the age of these assemblages through floral biostratigraphy.

We included Gondwanian assemblages characterized by *Nothor-hacopteris* from South America, Kashmir, and Australia in the late Visean–earliest Namurian A phytogeographic data set. Although Rigby (1973) felt these assemblages were Namurian–Westphalian in age in Australia and mostly Westphalian in South America, we feel *Nothorha-copteris* is certainly present in the Namurian A, and may be Visean in age, in Australia and South America for the following reasons:

1. Packham et al. (1969) reported *Rhacopteris* (*Nothorhacopter-is?*) in the Gilmore Volcanics of Australia, dated radiometrical-ly as Visean.
2. Crook (1961) reported a specimen of *Rhacopteris* (*Nothorha-copteris?*) from the top of the Parry Group, dated late Tournai-sian, in the Tamworth-Nundle District of New South Wales, Australia.
3. Frakes and Crowell (1969) placed the Tupe Formation at

Quebrada de la Herradura and Huaco Country in the Early Carboniferous (Strunian–Namurian C) on the basis of brachiopod stratigraphy. *Nothorhacopteris* occurs at both these localities (Archangelsky, 1971).

4. Gonzalez-Amicon (1973) correlated the palynoflora of the Retamito coal in San Juan, Argentina, with early Namurian palynofloras from the Lagares Formation in Argentina described by Menendez and Azucy (1969, 1971, 1973). *Nothorhacopteris* occurs associated with the Retamito deposits (Read, 1941; Archangelsky, 1971). Although some assemblages containing *Nothorhacopteris* are Westphalian in age (Morris, 1975; Rigby, 1973), evidence suggests that *Nothorhacopteris* may have appeared in the Visean.

We feel justified in regarding the *Nothorhacopteris* assemblages included in this analysis as late Visean–earliest Namurian A in age. However, the possibility exists that some late Namurian A–Westphalian assemblages have been included.

Different Habitats. Most compression floral assemblages form in wet lowland habitats such as swamps, levees, floodplains, and lake beds. Although these habitats often share species and genera, they may also contain genera and species that are limited to certain habitats. Because of this, including floral assemblages from different lowland habitats in a phytogeographic data set may cause differentiation on the basis of habitat rather than paleogeography. Ideally, a phytogeographic data set would include information on the habitat of each floral assemblage. In practice, this information is often unavailable in the published literature, making it virtually impossible to limit analyses to a specific habitat. Further, the small number of reliably dated assemblages often necessitates using every available assemblage to obtain an adequate sample size. For these reasons, habitat-related differentiation is virtually impossible to eliminate from phytogeographic analyses, although it may be minimized if the data collected from each paleogeographic region includes assemblages from the same habitats. However, habitat-related differentiation can mimic phytogeographic differentiation if sampling is poor and the assemblages collected from a paleogeographic region exclude a prevalent habitat.

On the other hand, the habitats that occur in a region may also contribute information about the paleoclimate and paleogeography. A region may lack swamp habitats because it has a dry paleoclimate or "upland" habitats because it has a wet paleoclimate. In our analysis of

Tournaisian–early Visean floral assemblages, both the scarcity of wet swampy habitats containing *Lepidodendron* lycopods in Acadia and the presence of Acadian endemics (*Diplotmema* and *Aneimites*) helped to differentiate these paleoclimatically dry assemblages from other tropical and subtropical assemblages. Under these circumstances, the elimination of habitat-related differentiation from paleogeographic data sets may be undesirable as well as impractical.

Sampling Intensity. An assemblage with low diversity may reflect the original diversity of the habitat, or it may reflect poor sampling. In general, floras from cold temperate paleolatitudes and dry paleolatitudes display lower diversity than floral assemblages from wet tropical and wet, warm temperate latitudes. Ideally a paleophytogeographic data set would exclude poorly sampled assemblages. However, there is no single standard for identifying floral assemblages in which poor sampling causes low diversity. Use of a rigorous diversity standard (that is, exclusion of all assemblages with fewer than four genera) might result in the exclusion of whole paleoclimatic zones from the analysis. Because of this, we included all floral assemblages with two or more vegetative compression taxa, which is the minimum number necessary for ordination analysis.

Standards for the Inclusion of Taxa

We chose the genus as the taxonomic unit for this study of Early Carboniferous phytogeography in order to avoid the problems of doing the study at the species level. Many compression fossil species have limited stratigraphic and paleogeographic distributions. Analysis at the generic level eliminates some of the problems posed by incorrect species-level identification and also enables us to use assemblages identified to the generic but not the specific level.

In correcting lists of the genera present in each floral assemblage, we eliminated certain genera on the basis of the following criteria:

1. genera designated "cf.";
2. genera found in only one locality;
3. genera identified on the basis of reproductive parts;
4. lycopod genera from the paleocontinent of Angara and not accepted by Meyen (1976, 1982) or Ananiev et al. (1979); and
5. lycopod genera from the paleocontinent of Gondwana and not accepted by Rigby (1969, 1973).

We discuss each of these criteria below.

Genera Designated "cf." We excluded genera designated "cf." because these fossils seemed the most likely to be misidentified. Exclusion of these forms helped to insure the most accurate data possible.

Single-Locality Endemics. We excluded genera that occurred in only one locality for two reasons. Multivariate statistical methods group assemblages on the basis of shared taxa, and genera that occur in only one locality add no information to such a study. In addition, recent work by Buzas et al. (1982) on the distribution of modern and fossil species suggests that a log series or lognormal distribution best describes the occurrence of species within localities. If this distribution also applies to genera, as many as 20 percent of the genera in our data set would be expected to occur in only one or two localities. Such rare taxa contribute little information to phytogeographic analyses. The presence of such taxa in a phytogeographic data set will only result in well-sampled localities appearing to have inflated levels of endemism.

Reproductive Genera. The reproductive and vegetative organs of plants often become sorted into different assemblages during deposition (Krasilov, 1975; Spicer, 1980). Because the presence of reproductive organs within a floral assemblage may depend on the mode of deposition, including such organs within an analysis introduces the possibility that floral assemblages derived from the same parent plants might differentiate on the basis of current velocity within the environment of deposition. To eliminate this source of nonphytogeographic differentiation in our data, we limited our study to vegetative organs; these occur more commonly and seem less susceptible than reproductive organs to current sorting during deposition.

Angaran Lycopods. Some reports of the compression floral assemblages from the Early Carboniferous of Angara (Siberia and Mongolia) did not contain detailed morphological descriptions of the taxa present (Durante, 1976). We relied on Ananiev et al. (1979) and Meyen (1976, 1982) as the arbitrators of the lycopod genera in the Early Carboniferous of the paleocontinent of Siberia. Thus we excluded *Prelepidodendron igryschens*, described by Durante (1976) in an assemblage from the Tourgen Ula Mountains of Mongolia, from the data used in this study. Meyen (1976) suggested that another lycopod assigned to *Prelepidodendron* (*P. varium*) actually belonged to the genus *Tomiodendron*.

When *Prelepidodendron* was excluded from the Mongolian assemblage, a species of this genus described by Danzé-Corsin (1960) from the Early Carboniferous of Morocco became a single-locality endemic and was also excluded from the data base.

Gondwanian Lycopods. We followed the convention of Rigby (1969) in regarding all Gondwanian occurrences of *Lepidodendron* as *Lepidodendropsis*.

Statistical Methods

We kept the individual floral lists separate for the purposes of the multivariate statistical analysis. Keeping the individual plant assemblages separate during the analysis has two advantages:

1. We can differentiate between rare and common endemic taxa within a region. A taxon that is endemic to a region and widespread within that region contains more information than an endemic taxon that is restricted to a few assemblages.
2. We decrease the effect that anomalous floral assemblages have on the floristic composition of the region as a whole.

Given m assemblages with n genera, all of the variation in the m assemblages could be expressed on an n-dimensional graph in which each axis represented one genus. Most people have difficulty understanding more than four or five dimensions at a time, and for a diverse data set this graph would be impossible to interpret. Multivariate statistical techniques proceed by trying to explain as much of the variation within a data set as possible on a number of axes that is small enough to comprehend.

We chose the multivariate statistical technique of polar ordination to analyze our data. Bray and Curtis (1957) introduced this technique to study the ecology of modern plants; Beals (1965), Whittaker (1978), Orloci (1978); and Cisne and Rabe (1978) presented detailed discussions of the applications and variations of this technique. We used the Fortran computer program POLAR II written by J. J. Sepkoski, Jr., and J. Sharry to produce the analyses presented in this study.

The result of a polar ordination analysis is a series of two-dimensional plots in which assemblages that share many genera plot close together and assemblages that share few genera plot far apart. For the polar ordinations in this study, we arranged the data into two matrices, one for the Tournaisian–early Visean (20 assemblages, 20 genera) and one for the late Visean–earliest Namurian A (35 assemblages, 37 genera). These matrices appear in tables 7A.2 and 7A.4 of the Appendix. We based our analyses on presence-absence data. Genera occurring in a locality were counted as 1; genera that did not occur were counted as 0. We would have preferred to use data that indicated the relative abundance of each genus in each assemblage; however, such data are impossible to obtain from published reports of Early Carboniferous compression floral assemblages.

Polar ordination begins with the calculation of a similarity matrix of assemblages based on the number of shared genera. We used two similarity indices in this analysis — the Dice index,

$$S_D = \frac{2a}{(n_1 + n_2)},$$

and the Otsuka index,

$$S_O = \frac{a}{(n_1 \cdot n_2)},$$

where n_1 = the number of genera in assemblage 1; n_2 = the number of genera in assemblage 2; and a = the number of genera shared between assemblage 1 and assemblage 2 (Cheetham and Hazel, 1969). The Otsuka index is equivalent to the cosine θ index when the cosine θ index is used with presence-absence data.

Both analyses produced similar results. However, analyses using the Dice similarity index yielded the most interpretable results and explained more of the variation in the similarity matrix. Because of this, we have presented only the results of analyses using the Dice similarity index.

In order to produce a polar ordination plot, the similarity matrix is converted to a dissimilarity matrix (one minus similarity). The sample that shows the greatest variance in dissimilarity relative to all other samples forms the endpoint of the first ordination axis. The sample that is most dissimilar to the chosen sample forms the opposite endpoint of the first ordination axis. The remaining assemblages are projected onto the first axis so that the distance between assemblages on the axes is proportional to their original dissimilarity to each other. The endpoints of the second axis are the assemblage that fits least well on axis 1 and the assemblage that is most dissimilar to that assemblage. The endpoints of the third axis are chosen in the same manner. The algorithm of axis choice constrains the three axes to be nearly perpendicular to each other and therefore independent.

The relationship between the variables (genera) and the ordination axes is given by the ordinated variable positions. The positions are calculated for each genus by summing the ordinated position of each assemblage containing the genus and dividing by the total number of assemblages. Genera that ordinate high on an axis characterize assemblages that ordinated high on that axis.

The correlation between the original distance matrix and the matrix of ordinated distances is called the cophenetic correlation (cf. Sneath and Sokal, 1973). The square of the value of the cophenetic correlation indicates the percentage of the variance in the original distance matrix retrieved or "explained" by the polar ordination analysis.

Polar ordination holds a distinct advantage over more traditional methods of phytogeographic analysis because it groups assemblages based on their total floral similarity, rather than the presence of a few geographically restricted taxa. This technique has had extensive use in investigations of distributional gradients among modern plants. It is ideally suited to uncover distributional gradients among fossil taxa, particularly during times with gentle latitudinal gradients or low generic diversity.

Results

Sphenopteris

In this study, we have assumed systematic names as a guide to the morphological similarity of fossils rather than to their phylogenetic relations, although we assume that morphology reflects both phylogeny and convergence. As long as vegetative morphology retains a consistent relationship to paleoclimate throughout the time interval under consideration, the inclusion of form genera in the data set should not obscure or create apparent differentiation within the data set. However, some plant compression taxa represent extremely generalized morphologies. For instance, plant compressions assigned to the genus *Sphenopteris* could represent the detached pinnules from the distal portions of *Diplotmema*, *Rhodea*, or *Sphenopteridium* (P. Gensel, pers. com., 1982). Because of this, *Sphenopteris* represents a broader range of leaf and branch morphologies than most of the other genera used. To insure that the inclusion of *Sphenopteris* did not create artificial phytogeographic patterns, we completed two Tournaisian–early Visean data analyses, one excluding and one including *Sphenopteris*. We found no significant differences between the results of the two analyses; that excluding *Sphenopteris* explained about 66 percent (total cophenetic correlation = 0.81) of the data present in the original correlation matrix; that including *Sphenopteris* explained 62 percent of the information present in the original correlation matrix (total cophenetic correlation =

0.79). If anything, *Sphenopteris* seems to contribute random or inexplicable variation to the data. In this paper, we discuss only the analyses that included *Sphenopteris* for two reasons. First, we wanted to remain as faithful as possible to the published literature; and second, we wanted to demonstrate the ability of these techniques to deal with data sets that contain a mixture of random, inexplicable variation and variation that can be explained by phytogeographic factors.

Tournaisian–Early Visean Phytogeography

The data for this analysis consisted of information about the presence or absence of 20 vegetative genera within 20 compression floral assemblages (see tables 7A.1 and 7A.2 of the Appendix). We completed Q-mode polar ordinations of the data using the Dice and Otsuka similarity indices. The best analysis (using the Dice similarity index) accounts for approximately 62 percent of the original variation present in the similarity matrix on three axes.

Figure 7.1 shows the localities of the Tournaisian and early Visean floral assemblages used in this study, plotted on a Tournaisian paleo-

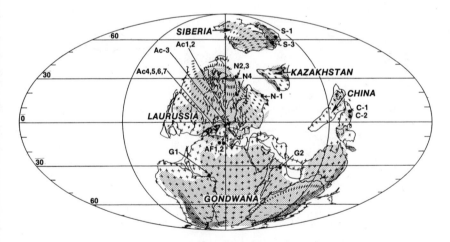

Fig. 7.1. Tournaisian–early Visean assemblages plotted on a Tournaisian paleogeographic reconstruction. See Appendix tables 7A.1 and 7A.2 for assemblage locations and composition. Paleocontinents are labeled. Assemblages marked S (Siberia), N (north low latitude), Ac (Acadia), E (equatorial), and G (Gondwana) belong to separate phytogeographic units. Assemblages marked AF (Africa) and C (China) belong to the same, separable phytogeographic unit.

geographic reconstruction. In this paper, we will refer to the latitudes 90–50° as high latitudes, the latitudes 50–30° as middle latitudes, the latitudes 30–15° as low latitudes, and the latitudes 15°N–15°S as equatorial latitudes.

Figure 7.2 shows a bivariate plot of the ordinated sample positions on the two "strongest" ordination axes (that is, those axes which individually returned the most variance). Axis 1 accounts for approximately 28 percent of the variance in the original similarity matrix. Axis 2 accounts for about 23 percent of the variance in the original similarity matrix. In this plot, axis 1 displays the separation between Mongolian floral assemblages from the paleocontinent of Siberia, which lay be-

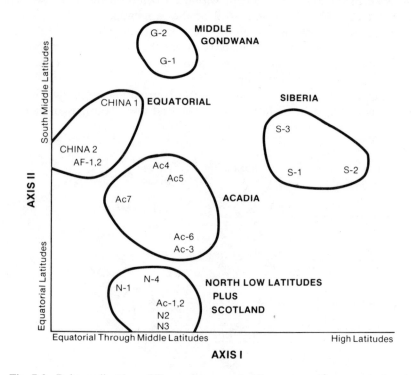

Fig. 7.2. Polar ordination of Tournaisian–early Visean assemblages, axes 1 and 2. Axis 1 separates assemblages on the paleocontinent of Siberia (S) from other Tournaisian–early Visean assemblages. Axis 2 reveals a gradient running from north low-latitude assemblages (N) to Acadian assemblages (Ac) to equatorial assemblages (E), to middle-Gondwanian assemblages (G). The Scottish assemblages (Ac-1, Ac-2) belong to the Acadian terrain but plot near north low-latitude assemblages on axes 1 and 2.

tween the paleolatitudes of 55°N and 65°N in the Tournaisian, and all other Tournaisian–early Visean floral assemblages, which lay between the paleolatitudes of 35°N and 40°S during these times.

Axis 2 displays a gradient among four groups of floral assemblages: (1) north low-latitude assemblages from Spitsbergen, Greenland, and the Urals; (2) Acadian assemblages from Scotland, Ireland, and the Maritime provinces, which lay between the paleolatitudes of 10°N and 0°; (3) equatorial assemblages from China and northern Africa; and (4) middle-Gondwanian assemblages from Ghana and Kashmir, which lay between the paleolatitudes of 30°S and 40°S. On axis 2, the two Scottish assemblages, which belong to the Acadian region, plot closer to north

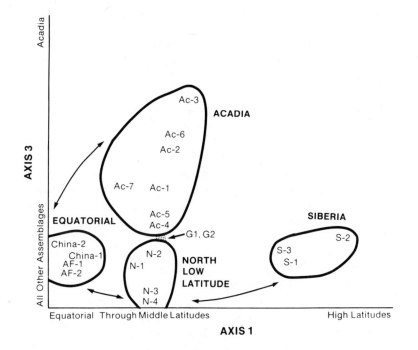

Fig. 7.3. Polar ordination of Tournaisian–early Visean assemblages, axes 1 and 3. Axis 1 separates Siberian assemblages (S) from other Tournaisian–early Visean assemblages. Axis 3 shows that Scottish assemblages (Ac-1, Ac-2), which plotted with north low-latitude assemblages (N) on axes 1 and 2, plot with other Acadian assemblages on axes 1 and 3. G-1 and G-2 represent middle-Gondwanian assemblages, which lie above Acadian and north low-latitude assemblages on axis 2.

low-latitude assemblages (the Urals, Spitsbergen, and Greenland) than to other Acadian assemblages. Both assemblages share *Lepidodendron* and *Stigmaria* with north low-latitude assemblages (see Appendix, table 7A.2). In addition, the assemblage from Foulden (Ac-2) lacks *Aneimites* and *Diplotmema*, which typically occur in the Tournaisian of Acadia. The presence of *Lepidodendron* and the absence of typical Acadian genera causes the Scottish assemblages to plot with the north low-latitude assemblages in figure 7.2.

Figure 7.3 shows a bivariate plot of floral assemblages on axes 1 and 3. Axis 3 accounts for about 15 percent of the variation in the original similarity matrix. As in figure 7.2, axis 1 separates Mongolian assemblages from other Tournaisian–early Visean assemblages. Axis 3 separates all the Acadian assemblages (including Scotland) from other Tournaisian–early Visean equatorial and low-latitude assemblages. Figure 7.3 shows that the Scottish assemblages that plotted near north low-latitude assemblages on axis 2 would lie above them in a three-dimensional plot showing axes 1, 2, and 3. In figure 7.3, middle-Gondwanian assemblages, which plot high on axis 2, fall in the middle of the plot, near Acadian and equatorial assemblages. In a three-dimensional plot, these assemblages would lie above the Acadian and equatorial and low-latitude assemblages.

Figures 7.2 and 7.3 show the separation among assemblages, while figures 7.4 and 7.5 show ordination plots of the genera important in determining the position of assemblages in figures 7.2 and 7.3. The position of a genus in figures 7.4 and 7.5 is calculated by summing the ordination position of each assemblage that contains the genus and dividing by the number of assemblages containing the genus. Genera that ordinate high on an axis characterize assemblages that ordinated high on that axis.

In figure 7.4, the genera *Tomiodendron*, *Pseudolepidodendron*, and *Ursodendron*, which characterize Mongolian assemblages, plot high on axis 1, near the position of Mongolian assemblages in figure 7.2. *Eolepidodendron*, *Triphyllopteris*, and *Asterocalamites*, genera that occur in equatorial and low-latitude assemblages, plot low on axis 1, near the position of these assemblages in figure 7.2.

Axis 2 in figure 7.2 separated north low-latitude assemblages (Greenland, Spitsbergen, and the Urals), Acadian assemblages, equatorial assemblages (China and North Africa), and middle-Gondwanian assemblages. In figure 7.4 *Stigmaria* and *Adiantites*, which characterized north low-latitude assemblages and Scottish assemblages, plot low on axis 2, near the position of these assemblages. The genus *Archeosigillaria*, which characterizes middle-Gondwanian assemblages, plots high on axis 2, near the position of middle-Gondwanian assemblages in figure 7.2.

The middle-Gondwanian assemblages, separated from other low-latitude and equatorial assemblages along axis 2, share the endemic lycopod genus *Archeosigillaria* but have poorly constrained ages (Mensah and Chaloner, 1971; Pal and Chaloner, 1979). The floral assemblage of Essipon, Ghana (G-1), is probably Early Carboniferous in age, but could be latest Devonian (Mensah and Chaloner, 1971). The floral assemblage from Kashmir (G-2) lies between a Tournaisian limestone and a marine shale of Upper Visean to Bashkirian (Namurian C–

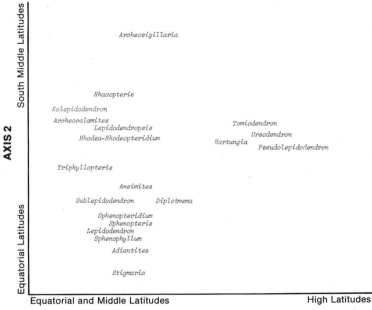

Fig. 7.4. Genera important for determining the positions of assemblages in fig. 7.2. The position of a genus is calculated by summing the ordination position of the assemblages that contain the genus and dividing by the number of assemblages that contain the genus. *Tomiodendron*, *Pseudolepidodendron*, and *Ursodendron* characterize Siberian assemblages. *Eolepidodendron*, *Triphyllopteris*, and *Archeocalamites* characterize equatorial and low-latitude assemblages. *Stigmaria* and *Adiantites* characterize north low-latitude (and Scottish) assemblages. *Archeosigillaria* characterizes middle-Gondwanian assemblages. See Appendix table 7A.2 for occurrences of each genus within the data set.

Westphalian A) age (Pal and Chaloner, 1979). Further study may prove this floral assemblage to be younger (Visean or Namurian) than the assemblage from Essipon, Ghana, to which we have compared it. If so, the relationship between *Rhacopteris* in the Kashmir floral assemblage and *Nothorhacopteris* in the Namurian of Gondwana may require further consideration. Due to the uncertainty of the ages of the middle-Gondwanian assemblages, the possibility remains that the differentiation of these assemblages reflects age as well as phytogeography.

Figure 7.5 shows the genera important in determining the position of assemblages in figure 7.3. As in figure 7.4, plants that characterize Mongolian assemblages plot high on axis 1 and plants that characterize north low-latitude assemblages plot low on axis 1. The plants that characterize Acadian assemblages (including Scottish assemblages), *Aneimites*, *Sphenopteris*, and *Diplotmema*, plot high on axis 3, in the position of the Acadian assemblages in figure 7.3.

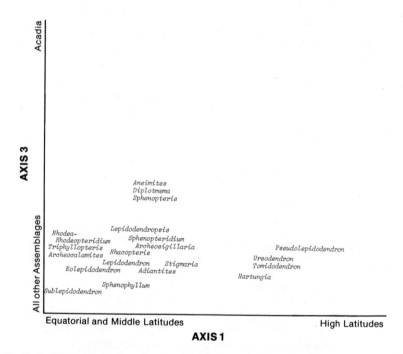

Fig. 7.5. Genera important for determining the position of assemblages in fig. 7.3. The positions of genera are determined as in fig. 7.4. *Aneimites*, *Sphenopteris*, and *Diplotmema* characterize Acadian assemblages, including Scotland. See Appendix table 7A.2 for occurrences of each genus within the data set.

Upper Visean–Earliest Namurian A Phytogeography

The data for this analysis consisted of information about the presence or absence of 37 vegetative genera in 35 floral localities (see tables 7A.3 and 7A.4 of the Appendix). Figure 7.6 shows these localities plotted on a Namurian A paleogeographic reconstruction. As with the Tournaisian–early Visean data, we completed Q-mode polar ordinations of the data using Dice and Otsuka similarity indices. In the best analysis (using the Dice similarity index), approximately 62 percent of the original variation present in the similarity matrix was retrieved on three axes.

Figure 7.7 shows a two-dimensional plot of the position of assemblages on polar ordination axes 1 and 2. Axis 1 accounts for approximately 45 percent and axis 2 accounts for approximately 16 percent of the variance in the original similarity matrix. Axis 1 differentiates Gondwanian assemblages from equatorial and north low-latitude assemblages and from Siberian and Mongolian assemblages. The Gondwanian assemblages occur in South America, Australia, and northern India between the paleolatitudes of 25°S and 60°S. The equatorial and north low-latitude assemblages occur in the western and central United States,

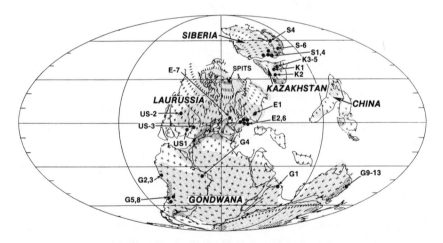

Fig. 7.6. Late Visean–earliest Namurian A assemblages plotted on a Namurian A reconstruction. Paleocontinents are labeled. On this reconstruction, northern Pangaea is labeled Laurussia; southern Pangaea is labeled Gondwana. See Appendix tables 7A.1 and 7A.2 for assemblage localities and composition. Assemblages labeled S (Siberia); assemblages labeled K (Kazakhstan), Spitz (Spitsbergen), E (Europe), and US (United States); and assemblages labeled G (Gondwana) belong to three separate phytogeographic units.

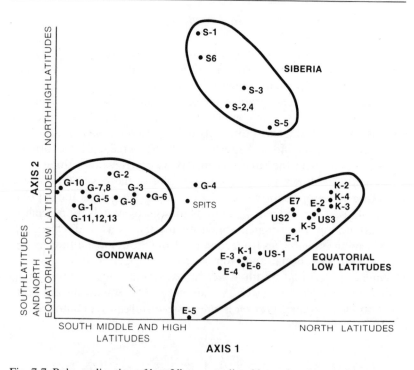

Fig. 7.7 Polar ordination of late Visean–earliest Namurian A assemblages, axes 1 and 2. Axis 1 separates Gondwanian assemblages (G) from Equatorial low-latitude assemblages (K, E, and US), and Siberian assemblages (S). Axis 2 separates assemblages from the paleocontinent of Siberia from Gondwanian and Equatorial low-latitude assemblages. Spitsbergen (Spits) and Brazil (G-4) plot in the middle of the diagram because they contain many widely distributed genera.

England, Europe, Spitsbergen, and Kazakhstan, between the paleolatitudes of 45°N and 5°S. The Angaran assemblages occur on the paleocontinent of Siberia, between the paleolatitudes of 65°N and 50°N. Axis 2 separates these Siberian and Mongolian assemblages from other groups of late Visean–earliest Namurian A assemblages.

The assemblages from Spitsbergen (Spits) and Brazil (G-4) plotted in the middle of figure 7.7, although we expected the Spitsbergen assemblage to plot with other equatorial and low-latitude assemblages and the Brazil assemblage to plot with those from Gondwana. In their comparison of ordination techniques, Gauch et al. (1977) found that samples that plotted in the middle of ordination axes were: (1) artificially constructed samples that contained randomly selected species; (2) samples composed mostly of species with widespread occurrences; or (3) samples that belonged in a central position along the environmental gradient. The assemblages from Brazil and Spitsbergen appear to plot in a median position because they contain many widely distributed genera: for example, *Adiantites*, *Rhodea*, *Sphenopteridium*, and *Sphenopteris* from equatorial and low latitudes, from Siberia, and from Gondwana and *Cardiopteris* from equatorial and low latitudes and from Siberia (See Appendix, table 7A.4).

Axis 3 explains only 7 percent of the variation in the original similarity matrix. Plots of the position of assemblages on axis 1 *versus* axis 3 and on axis 2 *versus* axis 3 provide very little additional information besides that contained in axes 1 and 2. Because of this, we neither reproduce nor discuss axis 3 here, except to note that this axis separated Kazakhstanian, North American, and the Spitsbergen assemblages from eastern European assemblages.

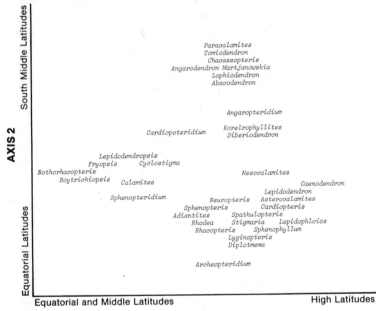

Fig. 7.8. Genera important for determining the position of assemblages in fig. 7.7. The position of a genus is determined as in fig. 7.4.
Nothorchacopteris characterizes Gondwanian assemblages.
Archeopteridium, *Caenodendron*, and *Lyginopteris* characterize equatorial low-latitude assemblages. *Paracalamites* and *Tomiodendron* characterize Siberian assemblages.

Figure 7.8 shows the ordination plot of the vegetative genera important in separating assemblages along axes 1 and 2. The genus *Nothorhacopteris*, which occurs in all the Gondwanian assemblages except Brazil (see Appendix, table 7A.4), plots low on axis 1; and *Caenodendron*, which characterizes assemblages from Kazakhstan, plots high on axis 1. *Paracalamites* and *Tomiodendron*, as well as other genera that characterize the assemblages from the paleocontinent of Siberia, plot high on axis 2. *Archeopteridium*, *Diplotmema*, and *Lyginopteris* characterize low-latitude assemblages and plot low on axis 2.

Discussion

Tournaisian–Early Visean Phytogeography

General. The separation between Mongolian floral assemblages and all other Tournaisian–early Visean floral assemblages encompasses most of the variation in this data set (see figs. 7.2 and 7.3). From a paleogeographic and paleoclimatic standpoint, this separation makes sense, as the Mongolian floral assemblages lay on a separate paleocontinent in the north high latitudes. Most of the other assemblages lay within the low and equatorial latitudes. We expect that the Mongolian assemblages experienced a seasonal climate with long, although not necessarily harsh, winters. Not surprisingly, most studies using traditional phytogeographic methods (Vakhrameev et al., 1970; Chaloner and Meyen, 1973; Chaloner and Lacey, 1973; Meyen, 1976, 1982) recognized this separation between Mongolian and Siberian (Angaran) assemblages and other Early Carboniferous assemblages.

Acadia. Phytogeographic analysis reveals a gradient among Acadian, equatorial, and north low-latitude assemblages. The separation of most of the Acadian assemblages (located between the paleolatitudes 10°S and 0°) from other equatorial assemblages (located between the paleolatitudes of 10°N and 10°S) could possibly be due to four factors: (1) the generic concepts of the workers who described the assemblages (Bell, 1948, 1960; Scott, 1979; Feehan, 1979); (2) east-west differentiation within the Tournaisian tropics; (3) the paleogeographic history of Acadia; (4) the paleoclimate of Acadia.

The differentiation of the Acadian assemblages probably does not result from the idiosyncratic taxonomy employed by the workers who described them. The descriptions of Acadian assemblages span 32 years and were completed by workers on two continents (North America and Europe).

Likewise, east-west phytogeographic differentiation within the Tournaisian–early Visean equatorial latitudes probably does not cause this differentiation. The vegetative genera of the Acadian assemblages are quite different from those of the Pocono and Price Formations, which also lay in the western tropics of the Tournaisian (Read, 1955; Scheckler, 1979; Jennings, 1979; Appendix, tables 7A.1 and 7A.2), although according to Gensel and Skog (1977), both the Acadian terrain and the Pocono and Price assemblages share the seed genus *Lagenospermum*. If new work indicates greater floral similarity between these assemblages and Acadian assemblages, this would strengthen the case for east-west phytogeographic differentiation among Tournaisian assemblages.

Although the Acadian terrain does have a separate paleogeographic history from Laurussia prior to the Tournaisian (Scotese et al., 1984), this separate paleogeographic history does not appear to cause the observed differentiation. During the Early Devonian, Acadian and North American assemblages belong to the same phytogeographic unit despite their paleogeographic separation (Raymond, Parker, and Barrett, this volume). During this time, similar paleoclimates influenced the composition of the floral assemblages more than did their paleogeographic separation.

The paleoclimate of Acadia appears to cause the observed differentiation. Throughout the Devonian, the Acadian terrain was subjected to mountain building resulting from the collision of Acadia and Laurussia. Coarse clastics, oxidized sediments, and local evaporites indicate renewed uplift and semiarid climates within this terrain during the late Tournaisian (Howie and Barss, 1975; Howie, 1979). Local aridity caused by the rain shadow of these uplifted areas within Acadia may account for the phytogeographic differentiation of the Acadian assemblages. Except for the Scottish localities, the Acadian assemblages lack *Lepidodendron* and *Stigmaria*, which characterize wet equatorial assemblages throughout the Lower and Upper Carboniferous. The presence of *Lepidodendron* and *Stigmaria* causes one Scottish locality to group with other equatorial and north low-latitude assemblages and may indicate a wetter climate in the northern portions of the Acadian terrain. The paleoclimate of the Acadian terrain is discussed further in the section on Early Carboniferous paleoclimate.

Other Areas. The north low-latitude assemblages (Spitsbergen, Greenland, and the Urals) lay between 15°N and 35°N during Tournaisian times. Seasonal climate or generally cooler temperatures could account for the separation of these assemblages from the equatorial assemblages (China and Morocco), which lay between 10°N and 10°S.

Axis 2 in figure 7.2 displays the separation between middle-Gondwanian assemblages (Kashmir and Ghana) and other Tournaisian floral assemblages. Although a discrepancy in age may account for the differentiation of these assemblages, their separation from Tournaisian north low-latitude and equatorial assemblages seems consistent with Tournaisian paleogeography and the sampling of Gondwanian assemblages in this analysis.

These middle-Gondwanian assemblages lay on the boundary between south low latitudes and middle latitudes. A floristic gradient, similar to the one linking Tournaisian north low-latitude and equatorial assemblages, may have linked these middle-Gondwanian assemblages to northern Gondwanian assemblages from Morocco. However, there are very few published accounts of Gondwanian assemblages of this age. Due to the small sample number it is difficult to separate a sharply polarized pattern from a gradient. Among the published material, Mensah and Chaloner (1971) reported the presence of *Archeosigillaria* in the Lower Carboniferous of Nigeria and Danzé-Corsin (1960) reported the presence of "cf. *Archeosigillaria*" in the Visean of Morocco. We could not determine the age of the Nigerian assemblage and excluded the Moroccan assemblage from the analysis on the basis of age. However, this evidence suggests that *Archeosigillaria*, a middle-Gondwanian genus, also occurs in northern Africa during the Early Carboniferous. Data on other northern Gondwanian assemblages might have revealed a floristic gradient.

In summary, phytogeographic analysis of Tournaisian–early Visean floral assemblages revealed five phytogeographic units; Siberian, north low-latitude, Acadian, equatorial, and middle-Gondwanian. Three of the five units (north low-latitude, Acadian, and equatorial) occurred along a floristic gradient linking semiarid Acadian assemblages with wet equatorial assemblages and wet equatorial assemblages with north low-latitude assemblages possibly living under seasonal conditions. The Siberian unit was confined to the paleocontinent of Siberia in north high paleolatitudes. The middle-Gondwanian unit lay in south low to middle paleolatitudes and presumably grew in a seasonal climate. A distributional gradient may have linked this assemblage to northern Gondwanian equatorial assemblages in Morocco.

No assemblage plotted in a position distant from other assemblages from the same paleocontinent, terrain, or paleolatitude. Although the Scottish assemblages plotted nearest north low-latitude assemblages on axis 1 (fig. 7.2), they plotted with other Acadian assemblages on axis 3 (figure 7.3). The dual affiliation of the Scottish assemblages is consistent with gradational phytogeographic boundaries between Acadian, north low-latitude, equatorial, and middle-Gondwanian assemblages.

Late Visean–Earliest Namurian A Phytogeography

The separation between Gondwanian floral assemblages and other late Visean–earliest Namurian A assemblages encompasses most of the variation in the data set (see fig. 7.7). Although we concluded in the discussion above that the age of these Gondwanian assemblages was late Visean–earliest Namurian A, the possibility remains that some of these assemblages are younger than early Namurian A; however, the phytogeographic separation observed seems reasonable. Gondwana and Laurussia had long, separate paleogeographic histories prior to the end of the Carboniferous, and ample evidence suggests that the paleoclimate of these two continents differed during late Visean and earliest Namurian A times.

The Gondwanian assemblages lay in southern middle and high latitudes (25°S–60°S) and would have experienced long, cold winters. Many of the *Nothorhacopteris* assemblages occur either above or below glacial sediments (Keidel and Harrington, 1938; Read, 1955), providing additional evidence for harsh climate in the southern hemisphere at the end of the Early Carboniferous.

Axis 2 in figure 7.7 displays the separation between Siberian and Mongolian assemblages and equatorial and north low- and middle-latitude assemblages. The Siberian and Mongolian assemblages lay between the paleolatitudes of 65°N and 50°N. As with Tournaisian–earliest Visean Mongolian assemblages, we suspect that these assemblages experienced long, although not necessarily harsh, winters.

The late Visean–earliest Namurian A phytogeographic analysis differs from that of the Tournaisian–early Visean analysis in that there is very little paleolatitudinal differentiation among the equatorial and north low- and middle-latitude assemblages. In the late Visean–earliest Namurian A phytogeographic data set, these assemblages lay between the paleolatitudes of 45°N and 5°S. Three of the Kazakhstanian assemblages, which lay between the paleolatitudes of 30°N and 45°N, do plot near the Siberian and Mongolian assemblages. However, a fourth, well-sampled assemblage from Kazakhstan contains *Stigmaria* and *Lepidodendron* (vegetative genera that characterize wet equatorial and low-latitude assemblages) and plots nearest to assemblages from Europe and the United States that lay in equatorial paleolatitudes.

In summary, phytogeographic analysis of late Visean–earliest Namurian A assemblages revealed three phytogeographic units; a

Siberian north high-latitude unit (which includes assemblages from Mongolia); a north low- and middle-latitude and equatorial unit; and a Gondwanian south high-latitude unit. In contrast to the Tournaisian–early Visean phytogeographic analysis, this analysis revealed little consistent phytogeographic differentiation among the equatorial and north low- and middle-latitude assemblages. In this analysis, only two of the 35 assemblages, Spitsbergen and Brazil, plotted far from other assemblages that belonged to the same paleocontinent or paleolatitudinal zone.

Implications for Early Carboniferous Paleogeography

The Tournaisian–early Visean phytogeographic analysis included 20 floral assemblages and 20 genera, with an average of 4.5 genera per assemblage. The late Visean–earliest Namurian A analysis included 35 floral assemblages and 37 genera, with an average of 6.3 genera per assemblage. We expected to observe more phytogeographic differentiation among late Visean–earliest Namurian A assemblages from the equatorial and low latitudes than among Tournaisian–early Visean equatorial and low-latitude assemblages. However, we observed a floristic gradient during the less diverse, earlier time interval linking dry equatorial assemblages, wet equatorial assemblages, and low-latitude assemblages between 10°S and 30°N. During the more diverse, later time interval, no observable phytogeographic differentiation occurred between the paleolatitudes of 30°N and 5°S, and the only gradational differentiation existed among Kazakhstanian assemblages from between the paleolatitudes of 30°N and 45°N and other equatorial and north low-latitude assemblages.

Paleogeographic changes caused by the collision of Laurussia and Gondwana could account for the decreased differentiation between equatorial and north low-latitude assemblages in late Visean–earliest Namurian A times. The events and timing of this collision are difficult to reconstruct (Lefort and Van der Voo, 1981), but we can assume that when a land bridge formed between Laurussia and Gondwana as a result of the collision, warm currents would have been "pooled" in the portion of Tethys east of the land bridge. The effect of this land bridge would have been to ameliorate climates in eastern Laurussia, Kazakhstan, southern Siberia, northern Gondwana, and China.

However, establishing the timing and sequence of events leading to land-bridge formation is complicated by the possibility of two kinds of marine bodies between Laurussia and Gondwana: (1) a deep ocean basin such as that shown in Barrett (this volume, figs. 5.3 and 5.4); or

(2) a shallow epicontinental sea across northern Africa such as that shown in the Tournaisian reconstruction used in this study (fig. 7.1). A change in paleoclimate between Tournaisian and the late Visean–earliest Namurian A due to pooling of warm water in eastern Tethys requires the formation of land bridges across both these seaways.

Bridging the Ocean Basin. Some sort of land bridge probably formed across the ocean basin between Laurussia and Gondwana between the Late Devonian and the Late Carboniferous as the result of the collision between these two paleocontinents. Lefort and Van der Voo (1981) felt that the collision started after the Visean. Scotese et al. (1984) fixed the younger limit for the occurrence of the collision in the late Carboniferous. Michard et al. (1982) suggested that deformation of northern African sediments caused by the collision began in the Late Devonian and that most of the thrust faulting of these sediments took place in the late Visean and early in the Late Carboniferous. If land-bridge formation accompanied this thrust faulting, a land bridge could have formed as early as the late Visean.

Bridging the Epicontinental Sea. Estimating the exact timing of land-bridge formation between Laurussia and Gondwana is further complicated by the presence of an epicontinental sea stretching across northern Africa. Such a sea could have maintained a marine connection between eastern and western Tethys after closure of the ocean basin and would have prevented the pooling of warm water on the eastern side of Tethys. Marine paleobiogeography provides the best evidence for the presence or absence of marine connections between the east and west sides of Laurussia and Gondwana. Hill (1973) reports links between the coral faunas of the Mississippi Valley and Eurasia (which includes northwest Africa) during the Tournaisian, suggesting a marine connection. In the Visean, the two regions share fewer genera. By Namurian time, the North American fauna was largely endemic, and northwest African faunas belong to the Eurasian province (Hill, 1973), indicating a barrier west of northwest Africa and east of the Mississippi Valley. Hodson and Ramsbottom (1973) found evidence for a land barrier blocking migration of goniatite cephalopods between North America and northwest Europe beginning in the Namurian. They located this land barrier as stretching from Ireland at least to Czechoslovakia. A southern extension of this land barrier into northern Africa would have blocked marine connection between eastern and western Laurussia and Gondwana. Marine paleobiogeography suggests that a complete barrier was in place by the late Visean or early Namurian and that warm currents dominated the eastern Tethys at this time.

Tournaisian and Namurian Paleoclimatology

We evaluated the paleoclimatic effects of such a late Visean–early Namurian land barrier using atmospheric oceanic and rainfall models for Tournaisian and Namurian paleocontinental reconstructions. Paleoclimatological studies have tended to emphasize the zonal aspect of climatic patterns — that is, the overall distribution of temperature and precipitation belts parallel to latitude (Briden and Irving, 1964; Parrish, Ziegler, and Scotese, 1982). Briden and Irving (1964) plotted the distribution of evaporites and coals against paleolatitude in pole-to-pole histograms and showed that coals tended to be centered around the paleolatitudes of 55°N and 55°S and around the paleoequator, whereas evaporites tended to center around the paleolatitudes of 30°N and 30°S. The distribution of these rocks corresponds well with the present global distribution of rainfall, which is higher at the equator and at about 55° north and south and lower at about 30° north and south.

However, the distribution of neither temperature nor rainfall follows a strict latitudinal gradient. Indeed, climatic asymmetry across the continents is the rule rather than the exception. A comparison of the climate in Los Angeles, California, with the climate in its colatitudinal city, Atlanta, Georgia, illustrates this phenomenon. The task of paleoclimatologists is to predict the asymmetric variation of rainfall and temperature within these general zonal patterns.

Although the controls on rainfall are very complex, first-order rainfall patterns may be predicted from geography and global atmospheric circulation. The general principles of atmospheric circulation patterns are well understood, and ancient circulation patterns can be predicted using global paleogeographic reconstructions (Parrish, 1982). With this knowledge, it is possible to predict first-order paleo–rainfall patterns (Parrish, Ziegler, and Scotese, 1982).

Atmospheric and Oceanic Circulation

Generation of rainfall models for any paleogeographic reconstruction requires the generation of atmospheric and oceanic circulation models. Parrish (1982) described the full algorithm for constructing atmospheric circulation maps. The equator-to-poles temperature gradient and the rotation of the Earth combine to create a general circulation composed of the following wind patterns: westward-flowing winds between about

20° north and south, converging on the equator; eastward-flowing winds between about 35° and 50° north and south in each hemisphere; and westward-flowing winds between about 65° and 75° north and south in each hemisphere. A corresponding latitudinal pressure regime also exists, with low pressure (ascending air) at the equator and at about 60° north and south and high pressure (descending air) at about 30° north and south and at the poles. This zonal pattern of winds and pressure systems dominates global circulation and reflects the effects of the thermal contrast between the equator and the poles.

The thermal contrast between land and sea constitutes the other major component of the thermal regime on the Earth's surface, and it modifies the zonal pattern created by the equator-to-poles gradient and the rotation of the Earth. Where land-sea contrast is sufficient, the zonal pattern will be disrupted or locally intensified by large-scale cellular features, such as the high-pressure cells at low midlatitudes over the oceans.

Figures 7.9 and 7.10 show the predicted atmospheric circulation patterns of Tournaisian and Namurian time. The geographic change over this interval was minimal. Its major influence was on ocean circulation, which tracks atmospheric circulation but is more confined by the distribution of land. The establishment of a land barrier across the equatorial seaway between Laurussia and Gondwana completely blocked the flow of the equatorial current (compare Figures 7.11 and 7.12). The effect of this blockage would have been to deflect all of the equatorial water along the coastlines on the western side of Tethys, making those coasts warmer and wetter at higher latitudes in the Namurian than in the Tournaisian. This would have obscured the latitudinal effect on temperature in that region and might cause the observed absence of phytogeographic differentiation between north low- and middle-latitude and equatorial assemblages in the late Visean and earliest Namurian A.

Another major difference between the Tournaisian and Namurian was the widening of southern Tethys, which permitted a more normal gyral oceanic circulation in the Namurian. This difference would have dried out the part of Gondwana that is now western Australia and made the coastal region in the vicinity of the present-day northeastern Africa and Saudi Arabia wetter. The passage between Laurussia and Siberia also narrowed considerably during the Tournasian, further blurring the climatic differentiation in western Tethys by ending the isolation of Siberia from relatively warm Tethyan waters.

Finally, the rotation of Gondwana southward over the pole restricted southern ocean circulation in the Namurian. One likely effect of

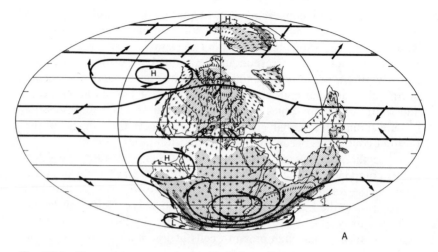

Fig. 7.9A. Tournaisian atmospheric circulation, northern summer. A
low-pressure cell would have been centered over Laurussia.
Southern Gondwana and the northwest coast of Gondwana would
have experienced high-pressure cells.

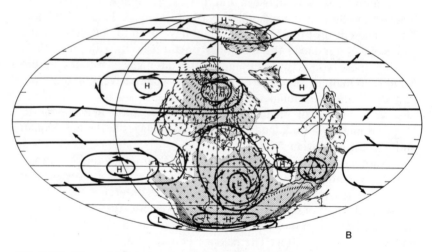

Fig. 7.9B. Tournaisian atmospheric circulation, northern winter. A
high-pressure cell would have been centered over Laurussia. Central
Gondwana and Australia would have experienced low-pressure cells.
A high-pressure cell would have been centered over southern
Gondwana.

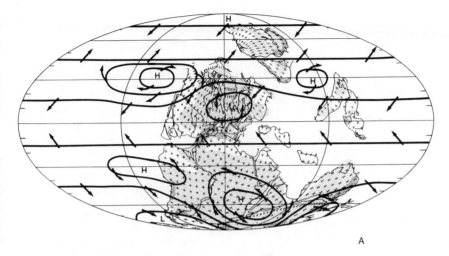

Fig. 7.10A. Namurian atmospheric circulation, northern summer. A low-pressure cell would have been centered over northern Pangaea (Laurussia). Central southern Pangaea (Gondwana) and the northwest coast of southern Pangaea (Gondwana) would have experienced high-pressure cells.

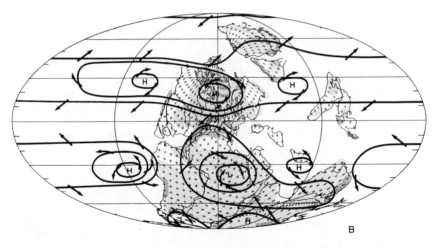

Fig. 7.10B. Namurian atmospheric circulation, northern winter. Northern Siberia and northern Pangaea (Laurussia) would have experienced low-pressure cells, as would central and northeastern southern Pangaea (Gondwana). The southernmost tip of Pangaea would have experienced a high-pressure cell.

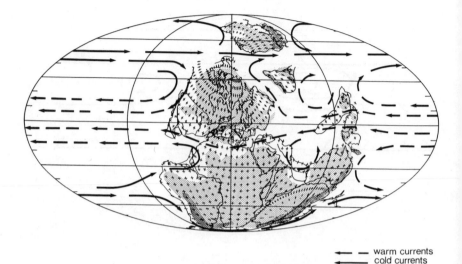

warm currents
cold currents

Fig. 7.11. Tournaisian oceanic circulation. The epicontinental sea in northern Gondwana would have allowed equatorial ocean currents to circle the globe, enhancing climatic differentiation along the east coast of Laurussia.

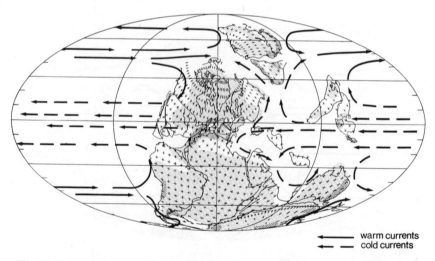

warm currents
cold currents

Fig. 7.12. Namurian oceanic circulation. A land bridge between northern and southern Pangaea (Laurussia and Gondwana) would have deflected equatorial currents north and south along the coast of Pangaea, ameliorating climate in northern Pangaea (Laurussia), Kazakhstan, and the southern tip of Siberia.

this movement would have been formation of a south polar ice cap. This in turn would have led to an equatorial compression of the terrestrial isotherms.

Rainfall. In constructing the rainfall maps (figs. 7.13 and 7.14) we considered the following factors: atmospheric pressure regime, wind direction, moisture source, and topography. The global latitudinal distribution of rainfall indicates the importance of the pressure regime. High rainfall tends to occur at the equator and at 55°N and 55°S, which are low-pressure areas. This happens because the rising air of a low-pressure zone cools, causing water vapor to condense and be released from the atmosphere as rain. However, the air flowing into a low-pressure zone must contain moisture in order to produce rain as it cools, so the source of the air is also important. An excellent example of the significance of the source of the air is the precipitation regime under the low-pressure cell that forms over the Great Basin of North America in the summer. The source for the air flowing into this low is the surrounding continental region, particularly the West. Much of the moisture in the western air is released as it passes over the Sierra Nevada. By the time the air ascends over the Great Basin, its moisture is already depleted and little rain is released. This example also illustrates

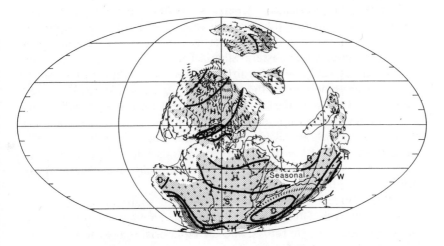

Fig. 7.13. Tournaisian rainfall. W = wet areas; H = humid areas; S = semiarid areas; D = dry areas. Eastern Laurussia, China, and northern Gondwana would have been wet, as would western Siberia. Northwestern Laurussia, northwestern Gondwana, and southeastern Gondwana would have been dry.

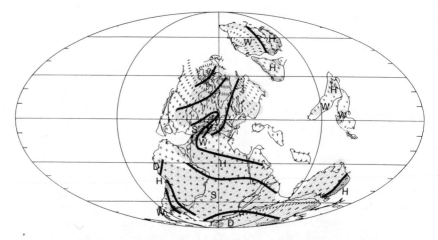

Fig. 7.14. Namurian rainfall. See 7.13 for key. Siberia, Kazakhstan, China, and
the east coast of Pangaea would have been wet or humid. The
northwestern and southern tips of Pangaea would have been dry.

the importance of topography — the windward sides of mountain ranges
are always wetter than the leeward sides unless the moisture in the air
passing over the mountains has already been depleted.

During the Tournaisian, Siberia would have been wet by virtue of
being in the temperate low-pressure zone. Moist, relatively warm air
from the southwest would have released most of its moisture over
western Siberia, while eastern Siberia (present-day Mongolia) would
have been less wet because the air would have been relatively less moist.

Moisture-laden air in the equatorial region would have released
some of its water over the small land areas of China, but the bulk of the
water would have been released over the parts of Laurussia straddling
the equator. The rainfall would have been particularly heavy on the
eastern sides of the Laurussian mountain ranges near the equator, and
the western portions of equatorial Laurussia would have been relatively
dry by virtue of being leeward.

The source of the air flowing into the southern summer low-
pressure cell over Gondwana would have been continental regions to
the south, the western Tethys, and the equatorial seaway to the north.
The northern coast of Gondwana, in the vicinity of what is now Saudi
Arabia, would have been wettest, but a humid area would have
extended inland, especially during the southern summer. The wet belt in
western Gondwana, in present-day South America, would have been
the southern equivalent of the wet region in Angara. Some of the

rainfall patterns discussed above, especially the ones in central and northern Gondwana, would have been seasonal. A Gondwanian desert, centered at 60°S, would have covered Antarctica.

The rainfall patterns in western South America, central and northern Gondwana, and Laurussia would have been much the same in the Namurian as they were in the Tournaisian. However, it is likely that more rain was released along the eastern coast of Laurussia, the southern coast of Siberia, and in Kazakhstan because the warm equatorial current would have been completely deflected by the land barrier between Laurussia and Gondwana, thereby increasing the influence of the warm ocean current in providing moisture to the coastal air.

The east coast of Australia would have been humid by virtue of the warm, moisture-laden equatorial air being deflected along the coastline around the oceanic high-pressure cell, especially in the southern summer. In addition, the Gondwanian desert would have covered the south pole in the Namurian.

Comparison of Phytogeographic Units and Paleoclimatic Models

The Tournaisian and Early Visean. The paleoclimatic models proposed for the Tournaisian agree well with the observed Tournaisian phytogeographic units. Floral assemblages from Mongolia lay in the eastern, humid portion of Siberia. The three groups of Tournaisian equatorial and low-latitude assemblages correspond to: equatorial humid and semiarid areas (Acadian assemblages); equatorial wet areas (North African and Chinese assemblages); and north low-latitude humid areas (assemblages from the Urals, Spitsbergen, and Greenland). The Acadian assemblages from Scotland lay near the northern edge of the Acadian terrain, on the border between semiarid and humid areas. Both of these assemblages contained *Lepidodendron* lycopods indicative of wet habitats, which is consistent with their occurrence on the border of a humid area. The middle-Gondwanian assemblages lay in the inland "humid summer" area of Gondwana.

The Late Visean and Earliest Namurian A. The oceanic circulation we have proposed for the late Visean and earliest Namurian A is similar to that of the Tethys sea during Permian times (Ziegler et al., 1981). During the Permian, deflection of warm equatorial currents along northern and southern coasts of Tethys allowed equatorial shelf faunas to flourish at anomalously high latitudes (Ziegler et al., 1981; Humphreville and Bambach, 1979). The presence of warm coastlines at relatively high latitudes during the late Visean and earliest Namurian A may account for the absence of phytogeographic differentiation between assemblages from north low and middle latitudes and equatorial paleolatitudes.

The late Visean–earliest Namurian A phytogeographic differentiation observed in the equatorial and low latitudes and in the southern hemisphere does not correspond closely to the proposed paleoclimatic model. Assemblages from the large equatorial–middle-latitude phytogeographic unit occur in the following paleoclimatic zones: north middle-latitude humid; north low-latitude humid; equatorial wet; and equatorial. Northward deflection of warm equatorial waters and the consequent amelioration of the pole-to-equator temperature gradient probably accounts for the lack of phytogeographic differentiation among these paleoclimatic zones. The results of the phytogeographic analyses presented here suggest that these paleoclimatic zones were less different from one another in the late Visean and earliest Namurian A than in the Tournaisian.

Assemblages from southern Gondwana fall into dry, humid, and wet rainfall zones. However, all these assemblages belong to the same phytogeographic unit. The reason for the lack of paleoclimatic differentiation among southern Gondwanian assemblages is not apparent.

Comparison with Palynofloral Phytogeography

At least two studies (Sullivan, 1967; Van der Zwan, 1981) addressed the palynofloral phytogeography of the Early Carboniferous. The results of Van der Zwan (1981), who treated middle Tournaisian–early Visean palynofloras, are somewhat difficult to compare to the results of our Tournaisian–earliest Visean analysis, primarily because Van der Zwan (1981) interpreted his results on a paleogeographic reconstruction that does not show the displacement of the Acadian terrains.

Van der Zwan (1981) described three phytogeographic regions based on palynomorphs: a "*frustulentus*" region, a *Vallatisporites* region, and a *Lophozonotriletes* region. The "*frustulentus*" region consists of Australian assemblages. The *Vallatisporites* region includes assemblages from the Acadian terrain as well as assemblage from Denmark. We modeled Denmark as wet equatorial and would have predicted this assemblage to group with assemblages from China and Morocco (see fig. 7.13). The *Lophozonotriletes* region consists of a north low-latitude assemblage from Spitsbergen. Van der Zwan considered assemblages from Europe (Russia and Poland), which we modeled as wet tropical, as transitional between the *Lophozonotriletes* region and the *Vallatisporites* region. He considered assemblages from China, which we placed in the wet tropics, as transitional between the *Vallatisporites* region (primarily Acadian) and the *"frustulentus"* region (Australia).

Van der Zwan (1981) observed the migration in the late Visean and early Namurian of taxa from the *Lophozonotriletes* region (north low-latitude) into the *Vallatisporites* region (primarily Acadian). He attributed this migration to northward drift of Eurasia (Laurussia); however, floral mixing between Acadian terrains and the north subtropics in the late Visean and early Namurian is consistent with the climatic changes accompanying the formation of a land bridge between Laurussia and Gondwana and the erosion of mountains connected to the Acadian terrains.

Sullivan (1967) described two palynofloral phytogeographic units, or suites, for the Tournaisian and three for the Visean. His Tournaisian *Vallatisporites* suite includes assemblages from the Acadian terrain, Ohio, Europe, Australia, and possibly northern Africa. Except for the inclusion of Australian assemblages, this palynophytogeographic unit includes assemblages from areas we modeled as dry equatorial and wet equatorial. The Tournaisian *Lophozonotriletes* suite of Sullivan (1967) includes assemblages from Spitsbergen, and from the Donets Basin of Russia, which has no compression floral assemblages of Tournaisian age. This suite seems to correspond to the north low-latitude unit described in this study, although on the rainfall maps we predicted that the Donets Basin would have had a wet equatorial palynoflora.

Turning to the Visean, Sullivan (1967) described three units. The *Grandispora* suite includes assemblages from the midwestern United States, northern Spain, eastern Europe, and Turkey. These areas lay between the paleolatitudes of 10°N and 15°S from the Tournaisian through the Namurian (figs. 7.1 and 7.6).

The *Monilospora* suite includes assemblages from Spitsbergen, the Donets Basin, and northern and western Canada. Except for the Donets Basin, which lay near the paleoequator, these areas lay between the paleolatitudes of 20°N and 35°N from the Tournaisian through the Namurian (fig. 7.1 and 7.6). The palynofloral phytogeographic studies of both Van der Zwan (1981) and Sullivan (1967) recognize a similarity between assemblages from the Donets Basin and Spitsbergen during Tournaisian and Visean times. These studies suggest that the climate of the Donets Basin, modeled here as wet equatorial, could have been cooler or drier and more similar to the climate of Spitsbergen during the Early Carboniferous.

The Visean Kazakhstanian suite of Sullivan (1967) has no macrofloral Tournaisian counterpart in this study, because we did not include any Kazakhstanian assemblages in our Tournaisian–early Visean data set. It also has no late Visean–earliest Namurian A counter-part, because the assemblages from Kazakhstan during this time interval grouped with north low-latitude and equatorial assemblages.

The results of Sullivan (1967) suggest that the great amount of differentiation that we observed in the Tournaisian continued into the Visean. Sullivan found evidence for three palynofloral provinces in the northern hemisphere during the Visean, two of which have Tournaisian macrofloral counterparts. The presence of a distinct Visean palynofloral unit on the palecontinent of Kazakhstan may indicate that a distinct Tournaisian macrofloral unit existed on Kazakhstan. If so, we would predict the presence of five Tournaisian phytogeographic units in the northern hemisphere. We would also predict that a seaway between Laurussia and Gondwana, which prevented the deflection of warm equatorial water along the coastline of western Tethys, persisted through most of the Visean prior to closure in the late Visean–earliest Namurian A.

However, it is important to remember that macrofloral and paly-nofloral units for the same interval may differ — particularly since the major control on macrofossil morphology, convergence due to paleocli-mate, may not affect palynomorph morphology to a great extent. In this case, phytogeographic units based on palynomorphs may come closer to "phylogenetic" biogeographic units than do units based on macrofossils.

Conclusions

Analysis of data from carefully delineated, relatively narrow time intervals suggests that five phytogeographic units existed in the Tournai-sian and early Visean and that three phytogeographic units existed in the late Visean and earliest Namurian A. These results are consistent with the wide paleolatitudinal separation of Early Carboniferous floral assemblages and previous studies of Early Carboniferous palynofloral phytogeography (Sullivan, 1967; Van der Zwan, 1981).

Phytogeographic differentiation decreased between Tournaisian and late Visean–earliest Namurian A times, although the number of nonendemic vegetative genera increased. Formation of a land bridge between terrestrial Laurussia and terrestrial Gondwana and the ensuing paleoclimatic changes may have caused this decrease in phytogeo-graphic differentiation. Paleoclimatic and rainfall models suggest that the results of such a land bridge would have been to ameliorate climate in north low latitudes and middle temperate paleolatitudes during the late Visean and earliest Namurian A.

The boundaries of paleoclimatic and rainfall zones in models of

Tournaisian climate correspond closely to the boundaries of Tournaisian phytogeographic units. Late Visean–earliest Namurian A paleoclimatic and rainfall zone boundaries do not correspond closely to the boundaries of the phytogeographic units delineated in this analysis. Such comparative studies can lead to a better understanding of paleoclimatic modeling and of the controls on phytogeographic differentiation.

Acknowledgments

The phytogeographic method introduced in this paper was developed during conversations with J. J. Sepkoski, Jr., W. F. Schmachtenberg, and S. F. Barrett, who also assisted in the translation of Russian and Portuguese articles. C. R. Scotese and A. M. Ziegler provided paleogeographic reconstructions and, along with S. F. Barrett and D. Rowley, provided insight into the biostratigraphy and taxonomy of Early Carboniferous compression assemblages. M. Fusco, A. Lottes, and A. O'Conner assisted in bibliographic research, data analysis, and translation of articles. We would like to thank S. Wing and R. Scott for their comments, and also J. Oliensis and B. Tiffney for their help in editing this article. K. McDonald and C. Lackey were tireless in their clerical efforts. Chevron Oil Company provided financial support.

Appendix

Table 7A.1   Floral Assemblages Used in Phytogeographic Analysis of the Tournaisian and Earliest Visean

Location	Abbreviation	Latitude	Longitude	Reference[*]
Bredy, Urals Mt., U.S.S.R.	N-1	52N	60E	1
Yudu, China	C-1	26N	115E	2
Jurong, China	C-2	32N	119E	3
Robert Tal, Spitsbergen	N-2	near 70N	20E	4,5
Camp Miller, Spitsbergen	N-3	near 70N	20E	4,5
Greenland	N-4	73N	23W	4
Djebel Bakach, Morocco†	AF-1	34N	6W	6
El Magnounan, Morocco	AF-2	34N	7W	6
Edrom, Scotland	Ac-1	56N	2W	7
Foulden, Scotland	Ac-2	56N	2W	7
Slieve Bloom, Ireland	Ac-3	53N	8W	8
Snake's Bight, Newfoundland	Ac-4	unknown		9
Albert Co., New Brunswick	Ac-5	45N	66W	10
Kennebecasis Is., New Brunswick	Ac-6	45N	66W	10
Norton Group, Nova Scotia	Ac-7	near 45N	62E	10
Essipon, Ghana	G-1	1N	12W	11
Gund Fm, Kashmir	G-2	34N	75W	12
Bay Daragin Gol, Mongolia	S-1	45N	99E	13
Eliegol, Mongolia	S-2	unknown		13,14
Tourgen Ula, Mongolia	S-3	50N	91E	13,14

[*] Reference numbers refer to the following articles: (1) Tchirkova, 1937; (2) Yang et al., 1979; (3) Zhao and Wu, 1979; (4) Hirmer, 1940; (5) Sveshnikova and Budantsev, 1969; (6) Danzé-Corsin, 1960; (7) Scott, 1979, and pers. com., 1981; (8) Feehan, 1979; (9) Bell, 1948; (10) Bell, 1960; (11) Mensah and Chaloner, 1971; (12) Pal and Chaloner, 1979; (13) Durante, 1976; (14) Ananiev et al., 1979.

† When *prelepidodendron* was excluded from the Mongolian assemblage, a species of this genus from Djebel Bakach became a single-locality endemic and was excluded from the data base.

Table 7A.2 Genera Counted as Present in Tournaisian–Earliest Visean Assemblages (1 = Presence)

Abbreviation	N-1	C-1	C-2	N-2	N-3	N-4	AF-1	AF-2	Ac-1	Ac-2	Ac-3	Ac-4	Ac-5	Ac-6	Ac-7	G-1	G-2	S-1	S-2	S-3
Adiantites	1											1	1	1	1					
Aneimites											1	1	1	1	1					
Archeosigillaria		1														1				
Archeocalamites		1	1				1			1	1	1	1		1	1				
Diplotmema		1	1																	
Eolepidodendron																				1
Hartungia																		1		
Lepidodendron	1	1	1	1	1		1	1	1	1	1	1	1	1	1					1
Lepidodendropsis		1	1				1	1	1	1	1	1	1		1					1
Pseudolepidodendron																			1	1
Rhacopteris							1					1				1	1			
Rhodea-Rhodeopteridium	1	1	1													1	1			
Sphenophyllum		1	1		1															
Sphenopteridium		1	1	1						1				1	1					
Sphenopteris	1	1		1				1	1	1	1			1	1					
Stigmaria	1	1		1	1			1	1	1										
Sublepidodendron	1	1		1	1		1	1	1											
Tomiodendron							1											1	1	1
Triphyllopteris														1					1	
Ursodendron															1				1	1

Table 7A.3 Floral Assemblages Used in Phytogeographic Analysis of the Late Visean and Earliest Namurian A

Location	Abbrevation	Latitude	Longitude	Reference[*]
Tomsk, Siberia	S-1	55N	85E	1
Minusinsk Basin, Siberia	S-2	54N	91E	1
Gorlovsk Basin, Siberia	S-3	55N	83E	1
Kuznetsk Basin, Siberia	S-4	54N	86E	1
Ara Teli Gol, Mongolia	S-5	49N	104E	2,3
Aimak Bayankongor, Mongolia	S-6	45N	99N	4
Karaganda Basin, Kazakhstan	K-1	56N	67E	5
Pri-Balkash, Kazakhstan	K-2	45N	75E	6
Dzungarian Region, Kazakhstan	K-3	45N	80E	6
Dzungarian Region, Kazakhstan	K-4	45N	80E	6
Dzungarian Region, Kazakhstan	K-5	45N	80E	6
Donnetz Basin, U.S.S.R.	E-1	49N	38E	7
Grätzer Graywacke, East Germany	E-2	51N	11E	8
Kyjovice Member, Czechoslovakia	E-3	50N	18E	8,9
Budisvice Member, Czechoslovakia	E-4	50N	18E	8
Malinowice Beds, Poland	E-5	50N	19E	10
Nova Ruda, Czechoslovakia	E-6	51N	16E	8
Millstone Grit, England	E-7	54N	3W	11

Location	Code	Latitude	Longitude	Ref.
Arkansas, U.S.A.	US-1	36N	94W	12
Utah, U.S.A.	US-2	40N	112W	13
Illinois, U.S.A.	US-3	38N	89W	14
Diabas Bucht, Spitsbergen	Spitz	near 70N	20E	8
Spitti, Kashmir	G-1	32N	78E	15
Carhuamayo, Peru	G-2	11S	76W	16
Paracas Peninsula, Peru	G-3	14S	76W	16
Teresina, Brazil	G-4	5S	43W	15
Retamito, Argentina	G-5	36S	69W	17
Quebrada de Tupe, Argentina	G-6	29S	66W	17
Quebrada de la Herradura, Argentina	G-7	26S	58W	17
Quebrada de los Cerros Bayos, Argentina	G-8	35S	68W	17
Clarke River, Australia	G-9	19S	146E	18
Stroud, Australia	G-10	32S	152E	18
Patterson, Australia	G-11	32S	152E	18
Port Stephen, Australia	G-12	33S	153E	18
Currabubula, Australia	G-13	31S	151E	18

* Reference numbers refer to the following articles: (1) Gorlova, 1972; (2) Ananiev et al., 1979; (3) Durante, 1976; (4) Daber, 1972; (5) Oshurkova, 1978; (6) Radchenko, 1967; (7) Novik, 1974; (8) Hirmer, 1940; (9) Purkynova, 1977; (10) Kotasowa, 1977; (11) Lacey, 1952; (12) White, 1937; (13) Arnold and Sadlick, 1962; (14) Lacey and Eggert, 1964; (15) Rigby, 1969; (16) Read, 1941; (17) Archangelsky, 1971; (18) Rigby, 1973.

Table 7A.4 Genera Counted as Present in Late Visean–Earliest Namurian A Assemblages (1 = Presence)

Abbreviation	S-1	S-2	S-3	S-4	S-5	S-6	K-1	K-2	K-3	K-4	K-5	E-1	E-2	E-3	E-4	E-5	E-6	E-7
Abacodendron		1																
Adiantites				1											1	1		
Angarodendron		1	1															
Angaropteridium		1	1			1												
Archeopteridium															1			
Archeocalamites							1	1	1	1	1	1	1	1	1	1	1	1
Boytrichiopsis																	1	
Caenodendron					1		1	1	1	1	1							
Calamites							1											1
Cardiopteridium	1	1	1												1			
Cardiopteris	1	1	1		1									1			1	
Chacassopteris	1					1						1						
Cyclostigma																		
Diplotmema							1							1				
Eleutherophyllum												1		1	1	1		
Fryopsis																		1
Korelrophyllites		1					1										1	
Lepidodendron				1									1	1	1		1	1
Lepidodendropsis								1	1	1							1	
Lepidophloios				1									1	1	1		1	
Lophiodendron		1																
Lyginopteris		1	1									1		1	1	1	1	1
Marjanowskia		1	1															
Mesocalamites		1	1					1	1			1	1	1			1	1
Neuropteris		1		1									1		1	1	1	
Paracalamites	1	1					1					1		1				
Nothorhacopteris														1			1	
Rhacopteris														1			1	
Rhodea-Rhodeopteridium												1		1	1	1	1	1
Siberiodendron		1										1						
Spathulopteris									1					1		1	1	
Sphenophyllum							1							1	1	1		
Sphenopteridium							1							1	1		1	
Sphenopteris							1											
Stigmaria										1								
Triphyllopteris			1											1	1		1	
Tomiodendron	1	1	1	1	1													

Table 7A.4 (Continued)

Abbreviation	US-1	US-2	US-3	Spits	G-1	G-2	G-3	G-4	G-5	G-6	G-7	G-8	G-9	G-10	G-11	G-12	G-13
Abacodendron	1																
Adiantites							1	1					1				
Angarodendron				1													
Angaropteridium																	
Archeopteridium																	
Archeocalamites	1		1														
Boytrichiopsis		1							1	1	1	1					
Caenodendron									1	1	1						
Calamites							1		1	1	1	1		1			
Cardiopteridium				1													
Chacassopteris		1															
Cyclostigma								1					1				
Diplotmema																	
Eleutherophyllum																	
Fryopsis									1	1							
Koretrophyllites	1	1		1													
Lepidodendron	1	1		1		1					1						
Lepidodendropsis											1	1					
Lepidophloios																	
Lophiodendron																	
Lyginopteris	1																
Martjanowskia																	
Mesocalamites										1							
Neuropteris	1																
Paracalamites						1											
Notorhacopteris	1				1				1		1	1		1	1		1
Rhacopteris	1	1			1			1									
Rhodea-Rhodeopteridium																	
Siberiodendron																	
Spathulopteris																	
Sphenophyllum									1		1	1		1	1	1	1
Sphenopteridium					1												
Sphenopteris	1			1	1												
Stigmaria	1		1	1		1	1	1		1							
Triphyllopteris						1		1								1	
Tomiodendron																	

References

Ananiev, V. A., et al. 1979. Biostratigraphy of the Lower Carboniferous of central Siberia. *Ninth International Congress of Carboniferous Stratigraphy and Geology (Abst.)*, p. 4.

Archangelsky, S. 1971. Las tafofloras del sistema Paganzo en la Republica Argentina. *An. Acad. Bras. Cienc.* 43 (Suplemento): 67–88.

———. 1983. *Nothorhacopteris*, a new generic name for some Carboniferous monopinnate fronds of Gondwanaland (= *Rhacopteris ovata* auct. and *Pseudohacopteris*, Rigby, 1973). *Rev. Palaeobot. Paly.* 38:157–72.

Arnold, C. A., and W. Sadlick. 1962. A Mississippian flora from northeastern Utah and its faunal and stratigraphic relations. *Contrib. Univ. Mich. Mus. Paleontol.* 17:241–61.

Bailey, I. W., and E. W. Sinnott. 1915. A botanical index of Cretaceous and Tertiary climates. *Science* 41:831–34.

———. 1916. The climatic distribution of certain types of angiosperm leaves. *Am. J. Bot.* 3:24–39.

Beals, E. W. 1965. Ordination of some corticolous cryptogamic communities in south-central Wisconsin. *Oikos* 16:1–8.

Bell, W. A. 1948. Early Carboniferous strata of St. Georges Bay Area, Newfoundland. *Geol. Surv. Pap. (Geol. Surv. Can.)* 10:1–45.

———. 1960. Mississippian Horton Group of type Windsor Horton District, Nova Scotia. *Mem. — Geol. Surv. Can.* 314:1–112.

Bray, J. R., and J. T. Curtis. 1957. An ordination of the upland forest communities of southern Wisconsin. *Ecol. Monogr.* 27:325–49.

Briden, J. C., and E. Irving. 1964. Paleolatitude spectra of sedimentary paleoclimatic indicators. In *Problems in paleoclimatology*, ed. A. E. M. Nairn, 199–224. London and New York: Interscience.

Bureau, M. E. 1914. Bassin de la Basse Loire. *Études des Gîtes Minéraux de la France, Ministère des Traveaux publics*, vol. 2, fasc. 2, pp. 1–417.

Buzas, M. A., C. F. Koch, S. J. Culver, and N. F. Sohl. 1982. On the distribution of species occurrence. *Paleobiology* 8:143–50.

Chaloner, W. G., and W. S. Lacey. 1973. The distribution of Late Paleozoic floras. In *Organisms and continents through time*, ed. N. F. Hughes, 271–89. Special Papers in Palaeontology, 12. London: Palaeotological Association.

Chaloner, W. G., and S. V. Meyen. 1973. Carboniferous and Permian floras of the northern continents. In *Atlas of paleobiogeography*, ed. A. Hallam, 169–86. Amsterdam: Elsevier Scientific Publishing Co.

Cheetham, A. H., and J. E. Hazel. 1969. Binary (presence-absence) similarity coefficients. *J. Paleontol.* 43:1130–36.

Cisne, J. L., and B. D. Rabe. 1978. Coenocorrelation: Gradient analysis of fossil communities and its application in stratigraphy. *Lethaia* 11:341–64.

Crick, R. E. 1980. Integration of paleobiogeography and paleogeography: Evidence from Arenigian nautiloid biogeography. *J. Paleontol.* 54:1218–36.

Crook, K. A. W. 1961. Stratigraphy of the Parry Group (Upper Devonian–

Lower Carboniferous), Tamworth-Nundle District, N.S.W. *J. Proc. R. Soc. N.S.W.* 94:190–206.

Daber, R. 1972. Abbildungen und Beschreibungen unterkarbonischer Pflanzenreste aus der Mongolischen Volksrepublik. *Paläontologische Abhandlungen, Abteilung B. Paläobotanik* 3, no. 5:867–85.

Danzé-Corsin, P. 1960. Sur les flores Viseennes du Maroc. *Bull. Soc. Géol. Fr.* 7:590–99.

Durante, M. V. 1976. The Carboniferous and Permian stratigraphy of Mongolia on the basis of paleobotanical data [in Russian]. *Trans., J. Sov.-Mong. Sci.-Res. Geol. Exped.*, vol. 19.

Feehan, J. 1979. Plants from the Upper Old Red Sandstone of Slieve Bloom, County Offaly, Erie. *Geol. Mag.* 116:403–04.

Frakes, L. A., and J. C. Crowell. 1969. Late Paleozoic Glaciation: I, South America. *Bull. Geol. Soc. Am.* 80:1007–42.

Gauch, H. G., Jr., R. H. Whittaker, and T. R. Wentworth. 1977. A comparative study of reciprocal averaging analysis and other ordination techniques. *J. Ecol.* 65:147–74.

Gensel, P. G. 1979. Aspects of the Lower Carboniferous flora of the Price Formation, southwestern Virginia: New information on some foliage taxa. *Ninth International Congress of Carboniferous Stratigraphy and Geology (Abst.)*, p. 69.

Gensel, P. G., and J. E. Skog. 1977. Two Early Mississippian seeds from the Price Formation of southwestern Virginia. *Brittonia* 29:332–51.

Gonzalez-Amicon, O. R. 1973. Microflora Carbonica de la localidad de Retamito, provincia de San Juan. *Ameghiniana* 10:1–35.

Gorlova, S. G. 1972. The flora and stratigraphy of the coal-bearing Carboniferous of middle Siberia. *Palaeontographica (Abt. B)* 165:53–77.

Hill, D. 1973. Lower Carboniferous corals. In *Atlas of paleobiogeography*, ed. A. Hallam, 133–42. Amsterdam: Elsevier Scientific Publishing Co.

Hirmer, M. 1940. Die Pflanzen des Karbon und Perm und ihre Stratigraphische Bedeutung. *Palaeontographica (Abt. B.)* 84:45–102.

Hodson, F., and W. H. C. Ramsbottom. 1973. The distribution of Lower Carboniferous goniatite faunas in relation to suggested continental reconstruction for the period. In *Organisms and continents through time*, ed. N. F. Hughes, 321–29. Special Papers in Palaeontology, 12. London: Palaeontological Association.

Howie, R. D. 1979. Carboniferous evaporites in Atlantic Canada. *Ninth International Congress of Carboniferous Stratigraphy and Geology (Abst.)*, pp. 93–94.

Howie, R. D., and M. S. Barss, 1975. Paleogeography and sedimentation in the Upper Paleozoic, eastern Canada. In *Canada's continental margins*, ed. C. J. Yorath, E. R. Parker, and D. J. Glass. *Can. Soc. Petroleum Geol. Mem.* 4:45–57.

Humphreville, R., and R. K. Bambach. 1979. Influence of geography, climate and ocean circulation on the pattern of generic diversity of brachiopods in the Permian. *Geol. Soc. Am. Abstr. Programs* 11:447.

Jennings, J. R. 1979. Plant megafossils from the Mauch Chunk Formation in eastern Pennsylvania. *Ninth International Congress of Carboniferous Stratigraphy and Geology (Abst.)*, pp. 99–100.

Jongmans, W. J. 1952. Some problems on Carboniferous stratigraphy. *Troisième Congrès de Stratigraphie et de Géologie du Carbonfere, Heerlen, 1951 — Compte Rendu* 1:295–306.

——. 1954. The Carboniferous flora of Peru. *Bull. Geol. Brit. Mus. Nat. Hist.* 2:191–223.

Keidel, J., and H. J. Harrington. 1938. On the discovery of Lower Carboniferous tillites in the Precordillera of San Juan, western Argentina. *Geol. Mag.* 75:103–29.

Kotasowa, A. 1977. Palaeobotanical evidence for the boundary between the Lower and Upper Carboniferous in the Upper Silesian Coal Basin. In *Symposium on carboniferous stratigraphy*, ed. V. M. Holub and R. H. Wagner, 429–31. Prague: Geological Survey.

Krasilov, V. A. 1975. *Paleoecology of terrestrial plants*. Trans. H. Hardin. New York: John Wiley and Sons.

Lacey, W. S. 1952. Additions to the Millstone Grit flora of Lancashire. *Troisième Congrès de Stratigraphie et de Géologie du Carbonifere, Heerlen, 1951. — Compte Rendu* 2:379–83.

Lacey, W. S., and D. A. Eggert. 1964. A flora from the Chester Series (Upper Mississippian) of southern Illinois. *Am. J. Bot.* 51:975–85.

Lefort, J. P., and R. Van der Voo. 1981. A kinematic model for the collision and complete suturing between Gondwanaland and Laurussia in the Carboniferous. *J. Geol.* 89:537–50.

Menendez, C. A., and C. L. Azcuy. 1969. Microflora Carbonica de la localidad de Paganzo, provincia de la Rioja, Parte I. *Ameghiniana* 6:77–97.

——. 1971. Microflora Carbonica de la localidad de Paganzo, provincia de la Rioja, Parte II. *Ameghiniana* 8:25–36.

——. 1973. Microflora Carbonica de la localidad de Paganzo, provincia de la Rioja, Parte III. *Ameghiniana* 10:51–71.

Mensah, M. K., and W. G. Chaloner. 1971. Lower Carboniferous lycopods from Ghana. *Palaeontology* 14:357–69.

Meyen, S. V. 1976. Carboniferous and Permian lepidophytes of Angaraland. *Palaeontographica (Abt. B)* 157:112–54.

——. 1982. The Carboniferous and Permian floras of Angaraland (a synthesis). *Biol. Mem.* 7:1–110.

Michard, A., A. Yazidi, F. Benziane, H. Hollard, and S. Willefert. 1982. Foreland thrusts and olistromes on the pre-Sahara margin of the Variscan orogen, Morocco. *Geology* 10:253–56.

Morris, N. 1975. The *Rhacopteris* flora in New South Wales. In *Papers from the Third Gondwana Symposium, Canberra, Australia, 1973*, ed. K. S. W. Campbell, 99–108. Canberra: Australian National Univ. Press.

Novik, E. O. 1974. *The Carboniferous flora of the European portion of the U.S.S.R.* [in Russian]. Kiev: Naukova Dumka.

Orloci, L. 1978. *Multivariate analysis in vegetation research*. The Hague: Junk.

Oshurkova, M. W. 1978. Paleophytocoenogenesis as the basis of a detailed stratigraphy with special reference to the Carboniferous of the Karaganda Basin. *Rev. Palaeobot. Palynol.* 25:181–87.

Packham, G. H., et al. 1969. The geology of New South Wales. *J. Geol. Soc. Aust.*, vol. 16, p. 1.

Pal, A. K., and W. G. Chaloner. 1979. A Lower Carboniferous *Lepidodendropsis* flora in Kashmir. *Nature* 281:295–97.

Parrish, J. T. 1982. Upwelling and petroleum source beds, with reference to the Paleozoic. *Am. Assoc. Pet. Geol. Bull.* 66:750–74.

Parrish, J. T., A. M. Ziegler, and C. R. Scotese. 1982. Rainfall patterns and the distribution of coals and evaporites in the Mesozoic and Cenozoic. *Palaeogeogr., Palaeoclimatol., Palaeoecol.* 40:67–101.

Purkynova, E. 1977. Namurian flora of the Moravian part of the Upper Silesian Coal Basin. In *Symposium on carboniferous stratigraphy*, ed. V. M. Holub and R. H. Wagner, 289–97. Prague: Geological Survey.

Radchenko, M. I. 1967. *Kamennougol'naia flora Iugo-Vostochnogo Kazakhstana.* Alma-Ata: Nauka Kazakhskoi SSR.

Read, C. B. 1941. Plantas Fosseis do Neo-Paleozoico do Parana E Santa Catarina. *Ministerio da Agricultura, Departmento Nacional da Producao Mineral, Divisao de Geologia E Mineralogia Monografia* 12:1–96.

———. 1955. Floras of the Pocono Formation and Price Sandstone in parts of Pennsylvania, Maryland, West Virginia and Virginia. *Prof. Pap. U.S. Geol. Surv.* 262:1–32.

Rigby, J. F. 1969. A reevaluation of the pre-Gondwana Carboniferous flora. *An. Acad. Bras. Cienc.* 41:393–413.

———. 1973. *Gondwanidium* and other similar Upper Paleozoic genera, and their stratigraphic significance. *Publ.—Geol. Surv. Queensl.* 350:1–10.

Robinson, P. L. 1973. Paleoclimatology and continental drift. In *Implications of continental drift to the earth sciences*, ed. D. H. Tarling and S. K. Runcorn, 1:449–76. London: Academic Press.

Scheckler, S. E. 1979. Persistence of the Devonian plant group Barinophytales into the basal Carboniferous of Virginia, U.S.A. *Ninth International Congress of Carboniferous Stratigraphy and Geology (Abst.)*, pp. 193–94.

Schimper, A. F. W. 1903. *Plant geography on a physiological basis.* Trans. W. R. Fisher; ed. P. Groom and I. B. Balfour. Oxford: Oxford Univ. Press.

Scotese, C. R., R. Van der Voo, P. Giles, and R. E. Johnson. 1984. Paleomagnetic results from the Carboniferous of Nova Scotia. In *Permo-Triassic continental configurations and pre-Permian plate tectonics*, ed. R. Van der Voo, J. Penado, and C. R. Scotese. AGU Geodynamics Series. Forthcoming.

Scott, A. C. 1979. The distribution of Lower Carboniferous floras in northern Britain. *Ninth International Congress of Carboniferous Stratigraphy and Geology (Abst.)*, pp. 198–99.

Smith, A. G., J. C. Briden, and G. E. Drewry. 1973. Phanerozoic world maps. In *Organisms and continents through time*, ed. N. F. Hughes, 1–42. Special Papers in Palaeontology, 12. London: Palaeontological Association.

Sneath, P. H. A., and R. R. Sokal. 1973. *Numerical taxonomy*. San Francisco: W. H. Freeman.

Spicer, R. A. 1980. The importance of depositional sorting to the biostratigraphy of plant megafossils. In *Biostratigraphy of fossil plants*, ed. D. L. Dilcher and T. N. Taylor, 171–84. Stroudsburg, PA: Dowden, Hutchinson and Ross.

Sullivan, H. J. 1967. Regional differences in Mississippian spore assemblages. *Rev. Palaeobot. Palynol.* 1:185–92.

Sveshnikova, I. N., and L. Y. Budantsev. 1969. *Florulae Fossils Arcticae* [in Russian]. Leningrad: Nauka Fillia Leningradensis.

Tchirkova, H. 1937. Contribution nouvelle a la flora Carbonifere inférieure du versant oriental de l'Oural. In *Problems of paleontology*, ed. A. Hartmann-Weinberg, L. M. Kretschetovitsch, and T. M. Kusmin. *Pub. Lab. Paleontol. Moscow Univ.* 2:207–31.

Vakhrameev, V. A., I. A. Dobruskina, E. D. Zaklinskaya, and S. V. Meyen. 1970. Paleozoic and Mesozoic floras of Eurasia and phytogeography of this time [in Russian]. *Trans. Geol. Inst. Acad. Sci. U.S.S.R.* 208:1–48.

Van der Zwan, C. J. 1981. Palynology, phytogeography and climate of the Lower Carboniferous. *Palaeogeogr., Palaeoclimatol., Palaeoecol.* 33:279–310.

Walter, H. 1973. *Vegetation of the earth in relation to climate and the ecophysiological conditions*. Trans. J. Wieser. New York: Springer-Verlag.

White, D. 1913. The fossil flora of West Virginia, Part II. In *The Flora of West Virginia*, 390–453. West Virginia Geological Survey.

———. 1937. Fossil flora of the Wedington Sandstone member of the Fayetteville Shale. *Prof. Pap. U.S. Geol. Surv.* 186B:13–40.

Whittaker, R. H., ed. 1978. *Ordination of plant communities*. The Hague: Junk.

Whittington, H. B., and C. B. Hughes. 1972. Ordovician geography and faunal provinces deduced from trilobite distribution. *Philos. Trans. R. Soc. London, Ser. B* 263:234–78.

Yang Jing-zhi, Wu Wang-shi, Zhang Lin-xin, and Liao Zhuo-ting. 1979. *Advances in the Carboniferous biostratigraphy of China*. Nanjing Institute of Geology and Palaeontology, Academia Sinica, Nanjing, China.

Zhao Xiu-hu and Wu Xiu-yuan. 1979. *Carboniferous macrofloras of south China*. Nanjing Institute of Geology and Palaeontology, Academia Sinica, Nanjing, China.

Ziegler, A. M., R. K. Bambach, J. T. Parrish, S. F. Barrett, E. H. Gierlowski, W. C. Parker, A. Raymond, and J. J. Sepkoski, Jr. 1981. Paleozoic biogeography and climatology. In *Paleobotany, paleoecology, and evolution*, ed. K. J. Niklas, 2:231–66. New York: Praeger.

WILLIAM A. DIMICHELE, TOM L. PHILLIPS,
AND RUSSEL A. PEPPERS

The Influence of Climate and Depositional Environment on the Distribution and Evolution of Pennsylvanian Coal-Swamp Plants

Introduction

Most evolutionary and ecological concepts of Late Carboniferous plants are based on fossils from wetland depositional environments of tropical Euramerica (Chaloner, 1958; Pfefferkorn, 1980). Although tropical wetlands and other lowland regions included a variety of habitats from floodplains to coal swamps, the sharpest ecological distinction was between peat-accumulating, coal-swamp environments and those with mineral-rich or clastic substrates. Coal-swamp environments were poor in nutrients, low in pH, poorly oxygenated, and under the strong influence of periodic standing fresh water. Plant remains preserved in coal balls as permineralized peat reflect the standing vegetation that formed the autochthonous or hypoautochthonous deposits and thus are less influenced by taphonomic factors than are compression assemblages. Clastic-substrate environments were usually more diverse floristically than were coal swamps and were edaphically much less uniform, although generally more nutrient-rich. Fossil plants preserved in these compression-impression assemblages were strongly influenced by taphonomic factors and yield considerably different morphological information than do anatomically preserved plants of coal-ball peats.

The kinds of factors that influenced the composition of plant communities in coal swamps include (1) expansion in abundance of previously uncommon plants present within the swamps; (2) migration into coal swamps of plants from clastic environments, which can be

223

deduced from an understanding of ecology and the distribution of groups through time; (3) regional or general extinction, which is biostratigraphically obvious from numerous sources; and (4) evolution, which occurred both inside and outside the coal swamps. Evolutionary changes, recognized as speciation, are by far the most difficult to deduce and document in detail solely on the basis of evidence from coal swamps.

Coal palynology and coal balls indicate that the vegetation of Pennsylvanian coal swamps varied both stratigraphically (temporally) and regionally (spatially) in taxonomic composition and relative dominance of the constituent plants. During most of the Pennsylvanian, the vegetation in stratigraphically consecutive coal swamps tended to vary slightly. These periods of relatively little change were interrupted by geologically brief periods of major change that occurred approximately synchronously over a widespread geographic region. Evidence from coal resources, sedimentology, and paleogeography suggests that changes in vegetation were mediated by climate, particularly by available moisture (Zangerl and Richardson, 1963; Peppers, 1979; Broadhurst, Simpson, and Hardy, 1980; Phillips, Shepard, and De-Maris, 1980; Phillips and Peppers, 1984). The ecological amplitudes of the dominant trees in coal swamps (Phillips and DiMichele, 1981; DiMichele and Phillips, n.d.) support the patterns of climatic change deduced from other sources of evidence.

The major changes in coal-swamp vegetation occurred on a Euramerican-wide scale at the onset of and during two intervals that have been interpreted as having been drier. These dry intervals were progressively more severe and led to an even more severe third pulse at the beginning of the Permian (Phillips and Peppers, 1984). The synchronous change in vegetation of coal swamps over such different basinal settings as the Western Coal Region and the Appalachian Coal Region, and the occurrence of synchronous extinctions across Euramerica, imply further that climate was a controlling factor.

The Influence on Coal-Swamp Vegetation of Climate and Geologic Factors

The Euramerican coal belt occupied an equatorial position and had a tropical climate during Pennsylvanian time (Ziegler et al., 1981; Schopf, 1973, 1975, 1979). The numerous coals and thick accumulations of sediment in cratonic basins indicate that temperature and moisture availability combined with favorable tectonic settings to allow the development of lowland swamps, many of which were of great areal extent (Wanless, et al., 1963; Wanless, Baroffio, and Trescott, 1969). Large accumulations of peat resulted in bituminous coal seams up to four meters in thickness. Phillips, Shepard and DeMaris (1980) and Phillips and Peppers (1984) have plotted the stratigraphic distribution of identified coal resources in the United States. Patterns in the increase and decrease of these coal resources coincide broadly with the timing of continental collisions and mountain building that occurred during the Pennsylvanian (Ziegler et al., 1981; Parrish, 1982). Phillips and co-workers presume that the relative abundance of coal generally reflects the amount of total moisture available to the lowland regions in which coal swamps formed, assuming that otherwise favorable depositional conditions existed. We presume that most swamps formed in freshwater areas, although some coal-swamp plants may have been tolerant of brackish water. Independent studies of sedimentological patterns, where available, support these deductions (Zangerl and Richardson, 1963; Broadhurst, Simpson, and Hardy, 1980), at least as inferred for the first drier interval.

The flora and vegetation of these Pennsylvanian tropical coal swamps were strongly controlled by availability of fresh water during the Pennsylvanian and thus represent an index for the prevailing conditions of wetness during the existence of a swamp. Consequently, through geologic time, changes in coal-swamp vegetation reflected long-term climatic trends in wetness during the Pennsylvanian Period. Because of the water-holding capacity of the peat substrate, coal-swamp floras should have been less sensitive to minor fluctuations in moisture availability than were clastic-substrate floras (clastic swamps and other lowland vegetation). Therefore, the plant composition of the coal balls should reflect the regional climate. We would expect coal-ball floras to be more sensitive indicators of climatic trends than are compression floras because coal swamps were ecologically more homogeneous and had a distinct vegetation that accumulated *in situ*. Complicating the influence of climatic change were widespread tectonic events and eustatic sea-level fluctuations that affected directly the nature and extent of coastal regions available for swamp development, with the result that much of the change we see may have been a consequence of repeated major marine transgressions or regressions. However, once peat-forming swamp vegetation evolved, climatic fluctuations would be one of the most important causes of change in the composition of the swamps through direct or indirect influence on migrations, extinctions, and evolution.

The Pennsylvanian Period can be divided into five broad climatic intervals (fig. 8.1) based on patterns in coal resource abundance and

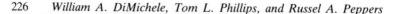

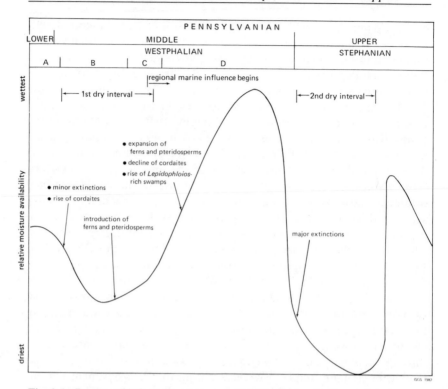

Fig. 8.1. Pattern of inferred moisture availability during the Pennsylvanian
          (after Phillips and Peppers, 1984).

corroborated by vegetational changes. (1) The Lower Pennsylvanian
(Westphalian A) was a time of moderate, probably seasonal wetness.
(2) During the early part of the Middle Pennsylvanian, a decline in
general moisture availability began, apparently throughout most of
Euramerica. This gradual decline was most severe during late Westphalian B time, after which a gradual increase in available moisture began.
(3) This increase of moisture, perhaps a consequence of greatly reduced
seasonality, culminated in the most favorable period for wetland
vegetation during the late Middle Pennsylvanian (middle to late Westphalian D), at least in what is now the United States. During this
increase in moisture, brackishness in coal swamps occurred in the
Illinois Basin Coal Field due to widespread marine transgressions that
began during the late Westphalian C and profoundly affected diversity
by regionally masking the effects of available fresh water. (4) An
apparently precipitous decline in availability of fresh water in mesic
lowland areas and swamps occurred at the boundary between the

Middle and Upper Pennsylvanian (Westphalian D–Stephanian A) in the Midcontinent region of the United States. (5) A return of climatic or tectonic conditions favorable for accumulation of thick peat deposits occurred during the second half of the Upper Pennsylvanian and was terminated by the onset of drier conditions in the Permian.

The drier intervals appear to have been progressively more severe. This is suggested by the successively greater disruption of coal-swamp communities, the more widespread nature of extinctions, and the rise to dominance of species that were progressively better adapted to cope with moisture stress. Coal swamps of the wetter intervals were relatively extensive and were often sites of very thick peat accumulation. These swamps were dominated by lycopods during the Lower and Middle Pennsylvanian and by tree ferns during the Upper Pennsylvanian. Coal swamps of the drier intervals were of limited areal extent and of variable thickness. Coal balls from coals of drier intervals are composed of a high percentage of roots, thus implying decay of aerial plant litter. The kinds of plants that dominated these swamps varied, but gymnosperms, particularly cordaites in the Middle Pennsylvanian, and tree ferns became significant parts of the vegetation at the expense of lycopods.

During periods of low moisture, coal swamps shrunk in size, became more isolated, thus local and regional variability of the flora became more common. A peat substrate presents an edaphic barrier to the migration of clastic-substrate plants onto peat substrates. This, in effect, makes coal swamps "closed systems"; that is, the development of local and regional variability of coal-swamp floras would affect the plant composition of later coal swamps, which were most likely to be populated by plants with coal-swamp ancestries. The spatial compression of vegetation confined to contracting swamps may have altered ecological relationships among species, caused local extinctions, and disrupted the genetic continuity of more extensive populations. One would predict that, as a consequence of the difficulty that many small populations would have had in maintaining continuity, these drier intervals were the most favorable either for the introduction of species into coal swamps from environments outside coal swamps or for the evolution and establishment of new species derived from coal-swamp ancestors. Smaller, drier coal swamps would have had a greater margin-to-area ratio, providing for more "edge effects" — that is, the length of ecotones would be increased relative to the area of the swamps isolated from contact with clastic-substrate vegetation. Thus, new species could be introduced in areas where resources were made available by disruption of the existing vegetation. The appearance of forms from the clastic-substrate areas also may be expected if swamps were drier, generally or seasonally, because there would be more areas of decaying, exposed peat substrate. This is suggested by the higher root-shoot ratios in coal balls deposited during a dry interval.

Coal Swamps as Evolutionary Islands

Most of the evolutionary inferences concerning Pennsylvanian plants have come in recent years from studies of coal balls. Coal balls provide some of the best-preserved plant material in the entire fossil record and morphological and anatomical information rivaling that available from extant plants. The major drawback in reconstructing phylogenies based solely on coal-ball plants is the strong possibility that these plants represent a biased sample of Pennsylvanian environments — only those portions of lineages that had evolved the capacity to contend with the unique edaphic constraints imposed by a coal swamp would be represented. Although coal-swamp vegetation may be a sensitive indicator of climatic change, it may be a less reliable source for direct phylogenetic reconstruction of most plant groups.

Edaphic factors related to a peat substrate are probably the most important general constraint on the kinds of plants that can exploit a peat-forming swamp. Modern peat swamps are environments of low nutrient availability for living plants (Schlesinger, 1978). Much organic material is incorporated into the substrate rather than being recycled. This organic matter chelates nutrient metallic ions, thereby restricting their availability to plants once these ions enter the substrate. The pH is usually low, and water stands in the swamp for at least part of the year. These conditions require that plants be able to cycle nutrients out of the water column, as well as have physiological and morphological characteristics allowing them to grow and reproduce under aquatic to semi-aquatic, low-nutrient, low-oxygen conditions. Clastic-substrate swamps may have many of the same edaphic constraints as peat swamps. However, greater influx of water, leading to nutrient replenishment and litter decay, make flooding regime the major factor controlling plant composition (Ehrenfeld and Gulick, 1981; Conner, Gosselink, and Parrondo, 1981), resulting in differences between clastic-substrate vegetation from flooded and nonflooded sites.

The marked differences in edaphic characteristics of peat and mineral substrates result in two general classes of environments with largely different constituent populations. Vegetation and flora in moist coal-swamp environments generally differed from vegetation in surrounding clastic-substrate areas. This suggests that coal swamps were relatively distinct centers of vegetational dynamics and evolutionary change. The unique edaphic properties of the coal swamps separated them as evolutionary islands from the more intergradational vegetation of clastic lowlands and clastic swamps.

These suppositions are borne out by recent quantitative studies of coal-swamp floras (Phillips et al., 1974; Phillips, Kunz, and Mickish, 1977; Eggert and Phillips, 1982; Phillips and DiMichele, 1981; DiMichele and Phillips, n.d.; Raymond and Phillips, 1983) and of clastic compression floras (Peppers and Pfefferkorn, 1970; Pfefferkorn, Mustafa, and Hass, 1975; Scott, 1977, 1978; Pfeiffer and Dilcher, 1979; Pfefferkorn and Thomson, 1982; Winston, 1983). Coal-swamp and clastic compression floras differ markedly in the relative abundances of major plant groups and, where representatives of groups can be found in both environments, in species composition. Floras growing on clastic substrates were diverse, as Scott (1977, 1978) has shown in detail; however, there was much greater taxonomic and vegetational similarity among clastic-substrate floras than there was between clastic and peat-swamp floras. Because attainment of the ability to grow in peat swamps was a more or less evolutionarily random process, some species of nearly every major group of Pennsylvanian plants evolved the capacity to exploit effectively some kind of coal-swamp setting, with some groups being much better represented than others.

The kinds of biotic interchange that occurred between peat-swamp and clastic-substrate environments varied. For some groups there was little commonality of species from the two general classes of substrate, suggesting subsequent discretely evolving lineages. In other groups, repeated introductions of species from clastic to peat substrates occurred. Introduced species tended to remain static in some groups, whereas in other cases they were stem species of lineages that evolved further *in situ*. Populations of some species may have bridged the edaphic lines between some parts of peat swamps and nonpeat areas. For the evolutionary biologist some questions of importance thus become (1) the degree of independence between coal-swamp and non-coal-swamp evolutionary patterns and (2) the degree with which one will reflect the other. Many species of some plant groups become major elements of coal-swamp floras, whereas only a few of the many genera and species of other groups were ever significant in coal-swamp habitats. Why this may have been the case provides a fertile area for paleoecological research.

Distribution of Major Plant Groups in Coal Swamps

Five major plant groups contributed significant biomass to coal-swamp peats during the Pennsylvanian Period. The groups were arborescent lycopods, cordaites, marattialean tree ferns, pteridosperms, and sphenopsids.

Certain lycopods were dominant or subdominant throughout the Lower and Middle Pennsylvanian, disappearing abruptly in North America with the onset of the second dry interval (see fig. 8.1). The genera and species abundant in coal swamps differed from those in clastic swamps, suggesting separate centers of evolution. The arborescent lycopods constitute the only plant group in which swamp-centered lineages can be recognized clearly.

Cordaitean gymnosperms probably were adapted to periodic or long-term moisture stress, caused either by actual low substrate moisture or by "physiological drought" from brackish-water conditions. Cordaites probably were centered in clastic lowland and upland areas (Chaloner, 1958) but moved into coal swamps during times of stress due to reduced moisture.

Tree ferns and medullosan pteridosperms were minor components of coal-swamp vegetation until the late Middle Pennsylvanian. Following their appearance in coal swamps, both groups became widespread, and their abundance characterized most coal swamps after many of the lycopods became extinct at the beginning of the second dry interval. Only a few species of tree ferns and pteridosperms were able to grow in the wettest coal-swamp habitats. Species of these groups apparently were introduced into coal swamps during the first dry interval, and some lineages subsequently adapted and expanded in importance as more moisture became available. It is possible that some of the species of tree ferns and pteridosperms that were relatively rare in coal swamps may have occupied clastic-substrate areas and marginal, higher-nutrient parts of coal swamps.

The sphenopsids, such as the calamites, were not typically coal-swamp plants but were most abundant on clastic substrates, especially in clastic swamps. Calamites may not have been represented by any uniquely coal-swamp-centered species, although *Sphenophyllum* may have undergone somewhat greater differentiation (Batenburg, 1982).

In the following discussion we consider the stratigraphic distribution of these major plant groups. Relationships of the distribution patterns to changes in climate aid in the assessment of whether the variation in morphology of coal-swamp plants reflects evolutionary continua or migrations from clastic substrates into coal swamps.

Lycopods

Lycopods were the major components of coal-swamp forests during the Lower and Middle Pennsylvanian. They can be divided into two major groups with different centers of distribution and evolution. One group, centered in coal swamps, included the genera *Lepidophloios, Lepidodendron* of the *L. vasculare* type, and *Paralycopodites*. The other group was centered in clastic lowland areas and included *Lepidodendron* of the *L. hickii* and *L. aculeatum* type, and *Sigillaria*; species of these genera were generally rare in coal swamps but became locally abundant in parts of swamps that may have been periodically drier or higher in nutrient levels. With the onset of the second dry interval in the midcontinent region of the United States the dominant kinds of coal-swamp lycopods became extinct, but some kinds of lycopods did survive.

It is necessary to distinguish two *Lepidodendron* groups representing two distinct ecological and evolutionary entities (genera). The name

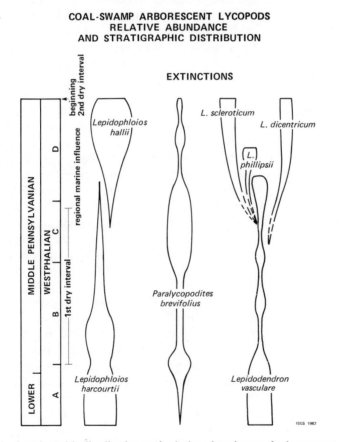

Fig. 8.2. Stratigraphic distribution and relative abundance of arborescent lycopods in Pennsylvanian coal swamps.

Lepidodendron is technically correct for the kind of arborescent lycopods centered in clastic environments, including *L. aculeatum*, which is
the type species of the genus (Thomas, 1970). True *Lepidodendron* is
represented in coal balls by *L. hickii* (DiMichele, 1983). Application of
the name *Lepidodendron* to *L. vasculare*, *L. scleroticum*, *L. phillipsii*,
and *L. dicentricum* encompasses the group of plants that paleobotanists
working with coal balls have come to treat as *Lepidodendron*
(DiMichele, 1981).

The lycopods, more than any other group of Pennsylvanian plants,
were particularly suited to coal-swamp habitats and show great stratigraphic continuity of species in such swamps. Several evolutionary
lineages of lycopods appear to have been centered in coal-swamp
environments, as suggested by the markedly different relative abundances of genera and species in clastic compressions and coal-ball floras.
The diversity of lycopods in coal swamps and our detailed knowledge of
their morphologies allow recognition of distinct patterns of distribution
and abundance changes of the major genera (fig. 8.2), patterns that
correlate generally with climatic fluctuations.

Lepidophloios. *Lepidophloios* was the most highly aquatically
adapted of the arborescent lycopods. The reproductive structures of
Lepidophloios (*Lepidocarpon*) and the general vegetative morphology
of the plant (Phillips, 1979; DiMichele and Phillips, n.d.; DiMichele,
1979a) reflect this aquatic adaptation. Community paleoecology indicates that *Lepidophloios* was strongly dominant (more than 70 percent
of biomass) in low-diversity communities having relatively few free-
sporing and ground-cover plants (Phillips and DiMichele, 1981). As a
consequence of extreme aquatic adaptations, and the abundance of
Lepidophloios in coal swamps, its distribution reflects more closely the
prevailing climatic wetness than does the distribution of the other
arborescent lycopods.

Lepidophloios harcourtii was relatively widespread and abundant
in Lower Pennsylvanian (Westphalian A) coal swamps of England and
elsewhere in western Europe during a time of moderately wet, monsoonal climate (Broadhurst, Simpson, and Hardy, 1980). There was a
general decline in its abundance at the onset of the first major dry
interval of the Pennsylvanian, a time from which it has been identified in
coals of eastern Kentucky, eastern Tennessee, and Iowa. The total
contribution of *L. harcourtii* to peat biomass during this period varies
greatly, based on the study of the few coals we have sampled. It appears
that the once rather extensive stands of *L. harcourtii* in the Lower
Pennsylvanian swamps may have been replaced by smaller populations

that were spatially and temporally isolated. During the late Middle Pennsylvanian *Lepidophloios*-dominated swamps increased in frequency and areal extent to become the most common kinds of coal swamps. Coinciding with this reexpansion of *Lepidophloios* swamps, *L. harcourtii* was replaced by *L. hallii* as the predominant form in coal basins of midcontinental United States (figs. 8.2 and 8.3). *Lepidophloios* began to decline just prior to the beginning of the second dry interval, at which time it became extinct.

A shift in the predominant species of *Lycospora* occurred during the first dry interval of the Pennsylvanian (Peppers, 1979; Phillips and Peppers, n.d.). *Lycospora* is the microspore of *Lepidophloios* and several other lycopod genera. A shift in dominance from *Lycospora pellucida* to *L. granulata* coincides with the *Lepidophloios harcourtii–L. hallii* shift. However, the considerable overlap of the two microspore species in space and time suggests that this was not an abrupt ecological replacement. It is possible that *L. hallii* evolved in coal swamps from an

ARBORESCENT LYCOPODS-RELATIVE ABUNDANCE AND STRATIGRAPHIC DISTRIBUTION IN COAL SWAMPS

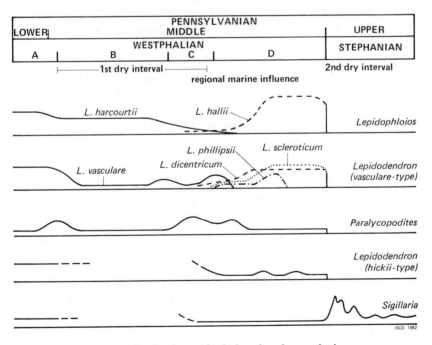

Fig. 8.3. Stratigraphic distribution and relative abundance of arborescent lycopods in Pennsylvanian coal swamps.

isolated population of *L. harcourtii*, incorporating the slight but consistent morphological differences between these two species (DiMichele, 1979a).

Lepidodendron (*L. vasculare* type). Four species of *Lepidodendron*, comprising two morphological groups, have been identified in coal balls. *Lepidodendron vasculare, L. scleroticum*, and *L. phillipsii* are morphologically similar and are distinct in several character states from *L. dicentricum*. The kind of morphological organization represented by these species is more common in coal balls from coal-swamp deposits than it is in compression floras from clastic-substrate areas. This may indicate that evolution in this genus was centered largely in coal swamps rather than in clastic swamps, although we cannot rule out introductions from clastic swamps as a source of changes in the composition of coal-swamp species through time. All four species produced *Achlamydocarpon varius* cones and *Capposporites distortus* microspores.

Although *Lepidodendron vasculare, L. scleroticum*, and *L. phillipsii* were morphologically similar, each species showed little intraspecific morphological variability. All produced small numbers of cones on deciduous lateral branches borne on an excurrent trunk (DiMichele, 1981). This provided them with the capacity to occupy sites subject to irregularly occurring, severe perturbations because reproduction was essentially continuous, providing a constant local background of propagules. Palynological studies (Peppers, 1979) indicate that *Capposporites* increased in abundance in areally discontinuous coal swamps or swamps that seem to have had peat deposition interrupted by transgressions or floods, regardless of the regional climatic trends.

During the Lower Pennsylvanian (Westphalian A) and up to the onset of the first dry interval, *Lepidodendron vasculare* was the only major species of the genus in coal swamps (fig. 8.2). According to megafossil samples, *Lepidodendron* was very rare in swamps during the drier period, although palynological studies indicate that it was locally abundant. During the period of increasing wetness, *L. vasculare* again became a major component of coal swamps. In the eastern regions of the Illinois Basin Coal Field, *L. vasculare* dominated small brackish swamps. It was a minor component in the Western Interior Coal Region (Iowa and Oklahoma). Occurring in conjunction with *L. vasculare* in some of these swamps were rare specimens of *L. phillipsii* and *L. scleroticum*, the earliest known appearances of these plants in coal balls. These species subsequently replaced *L. vasculare* as the areal extent of swamps and freshwater availability increased during the middle Westphalian D. *Lepidodendron phillipsii* reached maximum abundance locally in the Colchester (No. 2) and Summum (No. 4) Coal Members of Illinois, which palynology suggests were mostly dominated by *Lepidophloios* (Peppers, 1970). In these areas *Lepidodendron phillipsii* may have been growing in more elevated, perhaps drier regions of peat deposition. *Lepidodendron scleroticum* expanded as a component of *Lepidophloios hallii*–rich coal swamps. It is likely that *L. scleroticum* and *L. phillipsii* evolved from populations of *L. vasculare* during the first dry interval. These two species are similar to *L. vasculare* and may have displaced it as the areal extent of swamps increased in the late Middle Pennsylvanian (fig. 8.3). This pattern is similar to that suggested for *Lepidophloios*, but with more species involved.

Lepidodendron dicentricum trees had dendritic, determinate crowns with a large number of synchronously produced terminal cones (DiMichele, 1979b, 1981). The trees probably were spaced rather widely in coal swamps and were minor contributors to the biomass except where they occurred in assemblages of high diversity. This pattern is indicative of a fugitive ecological strategy. The species exhibits relatively high intraspecific morphological variability, which is a function of the presence of brackish and nonbrackish ecotypes (DiMichele, 1978, 1981).

The pattern of distribution of *Lepidodendron dicentricum* bears little relationship to those of the other species of *Lepidodendron*. Its earliest known stratigraphic occurrence is in eastern Tennessee in the middle Middle Pennsylvanian during a time of increasing wetness. *Lepidodendron dicentricum* occurs in small numbers until the time of major expansion of *Lepidophloios hallii*–rich swamps, in which it became a consistent but minor component. Stratigraphically, the brackish-water ecotype of *L. dicentricum* predominated early in the midcontinent swamps that were of relatively limited areal extent and proximate to marine conditions. As increase in the wetness and areal extent of swamps occurred, the brackish-water ecotype became restricted in distribution to the later phases of peat deposition in some swamps, while the freshwater form became more common. The later stages of these freshwater coal swamps may have been drier or under the influence of slightly brackish conditions prior to marine transgression.

Paralycopodites. *Paralycopodites* was an uncommon component of most coal-swamp vegetation. It became a major element in some swamps during the first dry interval and occurred sporadically after that time, prior to its extinction at the end of the Middle Pennsylvanian (fig. 8.2).

All specimens of *Paralycopodites* are morphologically very similar, and only a single species, *P. brevifolius* (DiMichele, 1980), is recognized. This small tree is associated with two species of cones in the Middle Pennsylvanian, *Flemingites schopfii* (Black, 1970) and *F. diversus* (Felix, 1954). *Flemingites schopfii* occurs in eastern Kentucky and differs in minor but consistent ways from *F. diversus*, which is found in the Eastern Interior Coal Province. This may indicate the existence of two (or more) species in coal swamps that otherwise did not differ detectably in vegetative morphology. Peat and palynological studies suggest that *Paralycopodites* was the major source of *Lycospora micropapillata*. Increases in the abundance of this spore correspond closely to increases in abundance of *Paralycopodites* in coal balls during the first dry interval. *Paralycopodites* grew largely in coal swamps, probably in areas of irregularly fluctuating low-water levels; the compression equivalent to *Paralycopodites* was probably *Ulodendron* (*sensu*, Thomas, 1967), a rarely encountered form.

Sigillaria and *Lepidodendron* (*L. hickii-aculeatum*). *Sigillaria* and true *Lepidodendron* were not centered in coal swamps. They are represented by numerous compression species that were common in a variety of clastic wetlands. The presence of species of these genera in coal swamps during the Lower and Middle Pennsylvanian suggests unusual conditions, perhaps involving increased clastic influx or seasonal dryness. Both genera are locally abundant in Lower Pennsylvanian (Westphalian A) coal balls of Europe. There are some indications from coal-ball and palynological data that these genera occurred locally in coal swamps during the first dry interval (fig. 8.3).

Lepidodendron hickii is a more significant coal-swamp element than *Sigillaria* in the Lower Pennsylvanian and in most Middle Pennsylvanian swamps where both occur. Occurrences in coal balls commonly are associated with local fluvial activity, as indicated by splitting of the coal by freshwater shales and sandstones. This suggests growth of *L. hickii* in rare nutrient-rich areas, possibly with great freshwater influx. The extinction of *L. hickii* and other species of true *Lepidodendron* at the end of the Middle Pennsylvanian in the midcontinent may be due to elimination of the clastic swamps in which they were most abundant. Such environments would have been affected by a regional decline in available moisture in much the same way as coal swamps. The almost complete, if temporary, elimination of the adaptive zone of coal- and clastic-swamp trees would have left few populations to recolonize swamps when favorable conditions returned and may be the key to extinction of such forms.

Species of *Sigillaria* occurred sporadically in Lower and Middle Pennsylvanian coal swamps, usually in association with indicators of drier coal-swamp conditions (Phillips and DiMichele, 1981). The survival of *Sigillaria* into the Upper Pennsylvanian of the midcontinent, despite the onset of the second, more severe, dry interval, may be due to the adaptation of many species to growth in mesic lowland areas that were not necessarily swamp habitats. Thus, a diversity of ecological tolerances among the species circumscribed a broad adaptive zone for the genus allowing persistence. *Sigillaria* is so uncommon in Middle Pennsylvanian coal-ball floras that species have not been delimited on the basis of anatomical data. It is not known whether Upper Pennsylvanian species of *Sigillaria* in coal swamps were carry-overs from Middle Pennsylvanian coal swamps or were later introductions from surrounding lowlands. Considering the lycopod extinctions at the second dry interval and the abundance of *Sigillaria* in compression floras, it seems probable that Upper Pennsylvanian species represent a reintroduction of the genus into coal swamps.

Cordaites

Cordaitean gymnosperms were important elements of coal-swamp vegetation during the first dry interval of the Pennsylvanian and during the period of regional brackish influence in midcontinent coal swamps that accompanied the subsequent increasing wetness (fig. 8.4). The group was able to enter coal swamps from surrounding lowlands during these times perhaps because of a general ability to grow under conditions of moisture stress (Wartmann, 1969). Two cordaitean genera were important in coal-swamp habitats — *Mesoxylon*, parent of *Mitrospermum* ovules, and *Pennsylvanioxylon*, which produced the ovule *Cardiocarpus*. Species of these genera probably had different tolerances to the parameters of moisture stress and brackishness. Basing their analysis on Iowa coal-ball floras, Raymond and Phillips (1983) suggested that the species of *Pennsylvanioxylon* that produced *Cardiocarpus spinatus* (and possibly one that produced *C. magnicellularis*) had the greatest tolerance of brackish conditions; *Mesoxylon*, although tolerant of decreased moisture availability, was not abundant in assemblages that were subject to marine influence or long-term brackish conditions. These conclusions, are supported by the stratigraphic distribution, the timing of appearances relative to climatic fluctuations, and the kinds of assemblages in which the species occurred. Another species of *Pennsylvanioxylon*, which produced *Cardiocarpus oviformis*, occurs in assemblages that appear to have been subjected to little or no brackishness,

CORDAITES-RELATIVE ABUNDANCE
AND STRATIGRAPHIC DISTRIBUTION IN COAL SWAMPS

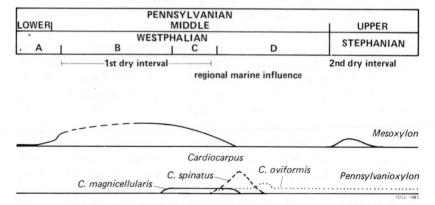

Fig. 8.4. Stratigraphic distribution and relative abundance of cordaites in
Pennsylvanian coal swamps.

particularly from coals of the upper Middle and Upper Pennsylvanian.

Mesoxylon was relatively rare in coal swamps prior to the first dry
interval. Coal balls in which *Mesoxylon* is abundant have high root-
shoot ratios and show evidence of extensive decay of aerial litter. This
suggests growth in those parts of coal swamps that may have been the
most periodically dry. *Mesoxylon* was replaced as the major cordaite in
coal swamps largely by *Pennsylvanioxylon* during the period of increasing wetness and concurrent brackish influence in the middle Middle
Pennsylvanian. *Mesoxylon* occurred again in some coal swamps of the
Upper Pennsylvanian during the second dry interval (fig. 8.4).

Pennsylvanioxylon first appeared in coal swamps during the first
dry interval (Westphalian B). The earliest ovules recorded are *Cardiocarpus leclercqii* (= *C. magnicellularis*) from coal k_8 of the Donets
Basin (Snigirevskaya, 1972) and the Aegir Horizon (Westphalian B-C
boundary) of the Netherlands (Koopmans, 1934). However, it was not
until later in the Middle Pennsylvanian that a *Pennsylvanioxylon*
species, producing *C. spinatus*, reached its peak abundance in coal
swamps that were subject to some brackish influence. These plants have
been interpreted as small mangrovelike trees (Cridland, 1964; Costanza, 1981). This abundance was short-lived, and the *C. spinatus* producers declined rapidly with the advent of extensive freshwater swamps.
The *Pennsylvanioxylon* that produced *Cardiocarpus oviformis* continued to be present in coal swamps through the upper Middle and

Upper Pennsylvanian. It was relatively abundant in parts of the Colchester (No. 2) and Summum (No. 4) Coals of Illinois, where *Lepidodendron phillipsii* was dominant, areas that presumably had the greatest elevation and fluctuations in water level in the swamps. Rothwell and Warner (n.d.) have reconstructed the *C. oviformis* producer is a small, shrubby plant in the Upper Pennsylvanian–age Duquesne Coal of Ohio.

The distribution of cordaites and the major areas of cordaite evolution are unresolved. The morphology of the plants suggests that they were xerophytes (Wartmann, 1969), and paleoecological studies support this conclusion (Peppers and Pfefferkorn, 1970; Scott, 1977; Raymond and Phillips, 1983). However, it is not clear if diversity and major evolution in the group was centered in "upland" (nonbasinal) areas (Chaloner, 1958) or whether the group was actually very diverse and widespread, with both lowland and upland components, possibly representing independent evolutionary lineages. If cordaites were centered primarily in uplands and mesic lowlands, the record of these plants in coal balls contains some serious phylogenetic gaps. These gaps may be least serious for *Pennsylvanioxylon*, which was possibly a coal-swamp-centered form. The three species of *Cardiocarpus* found in coal swamps overlap stratigraphically and vary from each other in minor ways, and the morphology of the species even varies regionally, suggesting that morphological variation originated in the swamps.

Medullosan Pteridosperms

The dominance of medullosan pteridosperms in many compression floras throughout the Pennsylvanian indicates that they were centered in mesic lowlands; only rarely were medullosans dominant in coal-swamp floras. *Medullosa* was a minor component of early Westphalian A coal swamps of Europe. This small but consistent representation continued throughout the early Middle Pennsylvanian in coal swamps in Illinois and the Appalachian Basin during the first dry interval. With increasing wetness in the late Middle Pennsylvanian, *Medullosa* became more abundant in coal swamps, an increase that correlated closely with the expansion of *Lepidophloios hallii*–rich swamps. At the time of their most extensive occurrences in the Middle Pennsylvanian, medullosans accounted for 10–15 percent of coal-swamp peat biomass. However, they were locally even more abundant, and their distribution was probably very discontinuous in coal swamps (Phillips, Kunz, and Mickish, 1977; Phillips and DiMichele, 1981). Medullosans became second in importance to tree ferns during the Upper Pennsylvanian.

Foliage is the most abundant kind of medullosan organ preserved and indicates, in a general manner, the amount of taxonomic overlap between coal-swamp and compression forms. All the species of *Neuropteris* or *Alethopteris* that have been identified in coal balls also occur in compression floras. These include *Neuropteris rarinervis* (Oestry-Stidd, 1979), *Neuropteris scheuchzeri* (Reihman and Schabilion, 1979), *Alethopteris lesquereuxii* (Baxter and Willhite, 1969), *Alethopteris sullivantii* (Leisman, 1960), and an *Alethopteris* much like *A. serli* (Mickle and Rothwell, 1982). *Mariopteris* (Stidd and Phillips, 1973), *Sphenopteris obtusiloba* (Shadle and Stidd, 1975) from the lyginopterid *Heterangium*, and *Linopteris* (Stidd, Oestry, and Phillips, 1975) also have been identified in both coal balls and compressions. In addition, there is a general overlap in size and shape of petrifaction and compression ovules. These features suggest that some of the same kinds of medullosans inhabited coal swamps as clastic-rich environments.

Medullosans probably grew in parts of coal swamps that were drier and higher in available nutrients than the lycopod-dominated parts of coal swamps. Quantitative studies of coal balls (Phillips, Kunz, and Mickish, 1977; Phillips and DiMichele, 1981) indicate that assemblages with more than 20 percent *Medullosa* are among the highest in fusain of any coal-swamp assemblages, indicating fires or peat oxidation and suggesting exposed substrates. Coal adjacent to coal-ball peat dominated by *Medullosa* is commonly high in mineral matter (Johnson, 1979), suggesting influx of clastics in floods or perhaps peat decay, which concentrates mineral matter.

The first dry interval fragmented and disrupted coal-swamp community structure, which may have rendered coal swamps more seasonally dry and edaphically heterogeneous. Under such conditions populations of medullosans could have established in coal-swamp communities more easily. With subsequent increased wetness and development of large, environmentally heterogeneous coal swamps, medullosan populations in coal swamps increased. Evidence from paleochannel environments suggests that levees supported larger stands of medullosans than generally were present in parts of coal swamps distant from channels. The overlap of coal-ball and compression foliage suggests that little intraswamp speciation occurred and that in some species there was overlap between clastic and coal-swamp populations.

Interpretation of the coal-ball record of the medullosans as a reliable, direct indication of phylogenetic patterns has great potential bias. If the establishment of medullosans in Middle Pennsylvanian coal swamps is primarily a product of random unidirectional introductions from mesic lowlands during periods of climatic disruption of swamp communities, then the coal-ball record is likely to represent a small, ecologically biased sample of the group. In this case, it would be difficult to track evolving lineages because of great incompleteness of the coal-ball record. On the other hand, if free exchanges occurred between clastic substrates and marginal, nutrient-rich or drier parts of coal swamps, then for some groups coal-ball floras may reflect general patterns of change. At present it appears that medullosan diversity in mesic lowland environments was much greater than that in coal swamps. Although some species were represented in both general classes of environments , coal swamps did not reflect total medullosan diversity. Thus, general evolutionary trends in the group may be deduced from coal balls, but it would appear that significant, ecologically mediated biases exist.

Marattialean Ferns

Marattialean ferns, like medullosan pteridosperms, are a morphologically and taxonomically complex and diverse group that occurs extensively in compressions and coal balls. Their pattern of distribution in coal swamps parallels that of the medullosans and suggests similar ecological mediation of distribution.

   Psaronius tree ferns occurred sporadically and in limited numbers in coal swamps during the Lower Pennsylvanian. According to palynological data, they diversified in regions outside coal swamps during the first dry interval in the early Middle Pennsylvanian, a diversification accompanied by introduction of a few additional forms to coal swamps (Peppers, 1979). *Psaronius* became a significant component of coal-swamp vegetation, as 5–10 percent of the peat biomass, during middle Middle Pennsylvanian time in the Illinois Basin. In the late Middle Pennsylvanian swamps *Psaronius* attained maximum abundance in widespread, freshwater *Lepidophloios*-rich coal swamps and in cordaite-dominated coal swamps subject to brackish influence. In Upper Pennsylvanian swamps *Psaronius* was usually the dominant element. At present it appears that few of the species important in Middle Pennsylvanian coal swamps persisted into the Upper Pennsylvanian.

   The best indications of marattialean taxonomic diversity are reproductive structures, which are difficult to correlate between compressions and petrifactions. Millay (1979, 1982) has approached this problem by comparing fertile pinnule morphology, the general aspect of sporangial placement, and palynology. He has found that many of the general groups of pecopterid foliage delimited by Corsin (1951) for compressions can be recognized in petrifactions, but that most coal-ball forms are similar to the *Pecopteris arborescens* group. Representation of the other groups is sparse and often rare. This scarcity suggests that a relatively small number of marattialean species accounts for most of the *Psaronius* biomass found in coal swamps. Quantitative studies of the Summum (No. 4) and Herrin (No. 6) Coals in the Illinois Basin indicate that, in these coals, most of the marattialeans are accounted for by the *Scolecopteris minor* group or *S. parvifolia* of Millay (1979); other species account for little of the total *Scolecopteris* biomass. Taxonomic studies indicate that marattialean diversity was generally low in any single coal swamp and that some species were restricted regionally.

Marattialeans are known to have had a long evolutionary history prior to their becoming major constituents in coal swamps. The earliest forms from the Upper Mississippian and Lower Pennsylvanian had simple monocyclic stelar architecture (DiMichele and Phillips, 1977). Increased anatomical complexity is first indicated by the presence of polycyclic stems that occur in nodules in shales above coals in the Westphalian A of England, suggesting that evolution of this anatomy occurred in mesic lowland habitats rather than in coal swamps. *Psaronius* was not a major part of coal-ball floras until the middle Middle Pennsylvanian; however, palynology provides insight into the diversification of marattialeans, which may have occurred during and immediately after the first dry interval (Peppers, 1979; Phillips and Peppers, 1984). The palynological data were derived from coals; however, coals of the first dry interval were areally restricted. Small coal swamps likely received abundant pollen and spore rain from adjacent areas and provide evidence of regional patterns of diversity. Thus, these coals may well reflect evolutionary diversification of this group in surrounding clastic wetlands rather than in coal swamps. If this were so, the later appearance of abundant *Psaronius* in coal-ball floras may reflect its subsequent invasion of coal swamps and expansion to levels detectable by coal-ball sampling.

Marattialean diversification and subsequent introduction into coal-swamp floras is difficult to interpret due to apparent conflicts in the lines of evidence. Data from coal palynology suggests a diversification in the early Middle Pennsylvanian; the tree ferns are presumed to be in, or marginal to, coal swamps. However, this is not corroborated by coal-ball studies or by analysis of compression-impression floras. According to megafossils marattialeans first became common in wetland areas during the late Middle Pennsylvanian (Pfefferkorn and Thomson, 1982; Phillips and Peppers, 1984). This suggests that the diversity of tree-fern spores in the early Middle Pennsylvanian coals was derived from regions in which preservation of megafossils was unlikely, possibly in lowland areas but in places with fully exposed soils. Taxonomic studies of fructifications show that many marattialean species occurred in coal swamps in general, but they also suggest that there was regional variability, with only a few species making a significant contribution to biomass. A few evolutionary lineages may have become centered in coal swamps after the radiations of the first dry interval. Species of these lineages had broad ecological amplitudes and were evolutionarily conservative. They persisted at least until the end of the Middle Pennsylvanian. Also occurring in the same swamps were less common species that were parts of lineages centered in clastic-wetland areas but that had the potential to occupy marginal parts of coal swamps.

Sphenopsids

Sphenopsids were represented in coal swamps by two major evolutionary groups, the arborescent calamites and the shrubby or viney *Sphenophyllum*. Both groups of sphenopsids occurred in coal swamps and were locally abundant or even dominant in clastic-substrate areas. Calamites are usually characterized as hydrophytes that grew along the margins of lakes or streams in shifting substrates and that were tolerant of periods of standing water (Scott, 1977, 1978). *Sphenophyllum* may have been more variable, growing in less frequently submerged areas (Batenburg, 1981). These deductions are supported by evidence from shale deposits in which calamite stems are often thickly layered or in which *Sphenophyllum* is preserved three-dimensionally through a considerable thickness of rock.

The presumed growth habit of both calamites and *Sphenophyllum*, spreading by the growth of rhizome systems, may explain their localized abundance in many compression floras. The calamites are the only group of Pennsylvanian arborescent lowland plants that had this potential for extensive vegetative propagation, which, like their modern counterpart, *Equisetum*, would have permitted them to colonize areas with unstable substrates. Their production of spore-bearing cones may have provided an effective means of dispersal to such colonizing plants occupying marginal, open-water areas.

Sphenophyllum was widespread and commonly occurred in swamps through most of the Pennsylvanian. Batenburg (1982) has suggested that there was an ecological differentiation between coal-swamp and clastic-lowland species, with little taxonomic overlap between the two general environments. Due to the small size of the plants, the species that grew in coal swamps were not significant contributors to the peat biomass. The calamites were also common throughout the Pennsylvanian in coal swamps, and despite their occasional large size (we have found stems at least 45 cm in diameter), their biomass contribution was low and confined to a few kinds of swamp assemblages often rich in *Medullosa* or *Sigillaria*. Palynology (Peppers, 1979) indicates an increase in abundance of calamites in some coal swamps, particularly during parts of the first dry interval, but this is exceptional. Fusinized preservation of calamite organs is common, and peat layers containing abundant calamite roots are frequently more common than those with abundant stems. Perhaps calamites were common in areas where decay of aerially exposed peat was pronounced, leaving few aerial remains that were commonly fusinized.

Calamites were probably centered outside coal swamps. They grew on loosely consolidated, shifting substrates, such as sand bars or stream and lake margins — habitats especially suited to rhizomatous plants. Where comparisons can be made, particularly on the basis of reproductive structures, it appears that there was a general overlap between coal-swamp and clastic-lowland forms. Only the genus that bore *Calamocarpon* (functionally monosporic megasporangia) was apparently persistent in coal swamps, but it was not very abundant.

Discussion

Pennsylvanian coal swamps were largely self-perpetuating, edaphically constrained habitats in the midst of the tropical lowlands; they were characterized by low taxonomic diversity and by the dominance of lower vascular plants. Although probably not important sites of evolutionary change (speciation), these swamps may have been evolutionary refugia for many lower vascular plant groups dependent on moist environments. Changes in vegetation during the Pennsylvanian were controlled in part by changing climatic patterns, particularly fluctuation in freshwater availability (net rainfall, runoff, and evapo-transpiration). Because of the "edaphic-island" nature of coal swamps, the changes in vegetation are excellent indicators of general trends in climatic conditions. However, edaphic factors excluded and many plants from coal swamps; thus, evidence from coal-swamp plants is an uncertain indication of the actual evolutionary patterns and morphological changes that accompanied individual speciation events within many phylogenies. Of the five major plant groups in coal swamps, the lycopods and ferns had the most endemic species, and it seems likely that only in the lycopods were some evolving lineages strongly centered in coal swamps.

The four major kinds of changes affecting the composition of coal-swamp vegetation during the Pennsylvanian (expansions, migrations and introductions, extinctions, and evolution) were most pronounced during times of marked climatic change. Quantitative shifts in the abundances of endemic coal-swamp trees are the major bases for paleoecological inferences about coal swamps. Frequently, the changes in the abundances of coal-swamp-centered species were coupled with the increasing abundance of newly introduced forms or new species. Migrations and introductions of plants into coal-swamp habitats from surrounding mesic or swampy lowlands probably account for most of the increased diversity in coal swamps during the Pennsylvanian. Introductions, which occurred during relatively short and defined periods, involved mainly the tree ferns, pteridosperms, and cordaites. These introductions coincided with times of inferred stress due to reduced moisture. The most obvious changes in composition of coal-swamp floras resulted from extinctions of major plant genera, which occurred during the Lower-Middle and Middle-Upper Pennsylvanian transitions. These extinctions included several arborescent lycopods and possibly many fern and pteridosperm species.

It is difficult to document evolution within coal-swamp ecosystems. It is often conjectural as to where, when, and under what ecological conditions new species appeared and to what degree their detection is affected by taphonomy. From an examination of changes in plant morphology in a given lineage through the entire Pennsylvanian, it is clear that evolutionary change took place. However, what makes phylogenetic inferences based on coal-swamp plants so precarious is the "island" nature of coal swamps, which preserve only parts of evolutionary lineages. Thus, morphological change observable within coal-swamp plants *per se* cannot be equated with direct lineal descent. Ancestor-descendant lineages are perhaps best represented in the lycopods. For most other plant groups the coal-swamp record may be one of isolated and random introductions of distantly related parts of evolving lineages that were centered outside coal-swamp habitats. Such a pattern of ecological distribution compounds the problem faced by paleontologists in reconstructing evolutionary (ancestor-descendant) relationships of organisms (Patterson, 1981), something that is inherently not falsifiable and that has received considerable discussion in the recent literature of systematic zoology.

Climate and Changes in Vegetation

The relative importance of the four major factors influencing the floristic composition of coal swamps differed during the two periods of vegetational change. As a result, these two periods of change represent two different kinds of events. During the first drier interval, introductions and subsequent quantitative shifts in abundance were most important and perhaps were linked to evolution of new taxa. Extinctions were of limited importance. The second dry interval was more severe; extinctions were pronounced and led to drastic quantitative shifts in composition of swamp floras. Subsequent introductions appear to have occurred in what were distinctly different kinds of coal swamps from those of the Middle Pennsylvanian. As indicated by the appearance of new species, the amount of evolution at this time was much greater in mesic lowland habitats than in coal swamps. However, our inability to discern evolutionary change within Upper Pennsylvanian coal swamps may also result in part from a lack of understanding of plants of this age. Palynology of coals in the Upper Pennsylvanian of the Illinois Basin has not been studied in as much detail as in the Middle Pennsylvanian.

First Dry Interval — Early Middle Pennsylvanian

The dry interval of the early Middle Pennsylvanian and the subsequent period of increasing moisture availability seem to be the major times during which evolutionary change occurred within Pennsylvanian coal swamps. It was in this period that maximum quantitative shifts in dominant and subdominant species occurred (but without loss of important genera, except for *Lyginopteris*) and that several groups appear to have been introduced into coal-swamp environments from the surrounding wetlands and mesic lowlands. The newly introduced groups rose to importance rapidly or expanded at a slightly later time as wetness and habitat heterogeneity in swamps increased. Cordaites were the first group to increase in abundance during the dry interval, followed by increases in tree ferns and then medullosan pteridosperms. All these groups were present to some extent in Lower Pennsylvanian coal swamps, and their increases may have involved expansion of endemic coal-swamp forms as well as introduction of species from surrounding lowlands.

The nature of the events leading to the floristic and vegetational changes of the Middle Pennsylvanian coal swamps is suggested as follows. (1) Decreased moisture availability, accentuated by seasonal dryness, resulted in contraction in area and time of duration of coal swamps. (2) Coal swamps were increasingly isolated from each other in the lowlands and were less edaphically distinctive from other wetlands; such coal swamps were "drier" with increased peat oxidation and decay.

(3) Swamp communities were progressively disrupted due to shrinking of physical settings conducive for coal-swamp development. Such swamps probably increased in heterogeneity, both spatially and temporally, accentuating the nonequilibrial aspects of community structure and resulting in incomplete utilization of available resources (Pickett, 1980).

These events would produce several different consequences. (1) The floristic composition of separate coal swamps during the dry interval should have been variable. (2) Plants from the surrounding lowlands, particularly clastic wetlands, may have been able to exploit coal-swamp resources because of the increased heterogeneity of coal-swamp habitats. (These previously had been excluded from the coal swamps because of the waterlogged soils and general environmental homogeneity, features that favored the establishment of low-diversity, "equilibrium" vegetation.) The coal swamps may have become mosaics of saturated and unsaturated substrates due to periodic drying of the peat, allowing some plants from clastic lowlands undergoing stress due to decreasing moisture to move into coal swamps. (3) For similar reasons, new species evolving in the lowlands, as well as in the swamps, would have had a greater chance of locating suitable habitats during these times of disruption of coal-swamp communities.

The genetic origin of variation is a continuously occurring process, providing the phenotypes that are the raw material of the morphological divergences we detect as speciation. However, one of the major constraints on the amount of speciation we detect is the ability of newly evolved lineages to survive and become established in the ecosystem. Barriers to establishment of new species in coal swamps should have been lowest during the first dry interval. If the species changes in coal swamps that we detect during this time period are due largely to establishment of newly evolved forms, we would find a pattern similar to that described by Williamson (1981) for mollusks in the Turkana Basin of East Africa — that is, proliferation of new morphotypes during times of increasing isolation and fragmentation of ancestral populations in combination with increasing environmental stress. The fossil record of the plants of this part of the Pennsylvanian is much less complete than that of Williamson's mollusks, and the ancestral species do not return to displace their descendants as climate moderates. However, the general similarities of mass morphological change correlated with a specific set of ecological changes is an interesting and noteworthy parallel.

At the end of the first dry interval, there would have been many locally established populations that had been introduced to coal swamps or that had expanded in abundance within coal swamps. These would have competed for increasing swamp resources. There also would have been fragmented populations of formerly dominant plants, such as the lycopods *Lepidophloios hallii* and *Lepidodendron vasculare*. These also may have given rise to new species populations due to long-term isolation and consequent genetic drift. Along with these plant populations were groups, such as the cordaites, that had risen to abundance during the drier interval because of their ability to cope with moisture stress.

As coal swamps became wetter and less seasonally limited and as available resources expanded, the competitive sorting-out of these species resulted in coal swamps with a different vegetation and flora than those that preceded the onset of dryness. The major swamp lycopods remained part of coal-swamp communities during the dry interval; with increasing wetness the earlier species were replaced by new but morphologically similar forms. Palynology indicates that fern diversity rose sharply during the driest part of the interval, when mesic lowlands were also under moisture stress. However, tree ferns do not appear in coal balls until the later Middle Pennsylvanian, as moisture availability increased. We lack the benefits of a precise palynological record of medullosan pteridosperms during the dry interval. Nevertheless, their later appearance in coal-swamp communities paralleled that of the ferns and suggests somewhat similar ecological constraints. The cordaitean genus *Mesoxylon* increased in abundance during the dry interval, a pattern consistent with the concept of cordaites as "upland"-centered plants adapted to drier sites. It declined in abundance with increasing wetness in the late Westphalian C.

The pattern of change in the transition from the first dry interval to more continuously wet conditions is complicated regionally by periods of influence of brackish water in midcontinental coal swamps. This brackish influence caused a short-term decline in diversity and a rise in brackish-tolerant plants, such as some species of *Pennsylvanioxylon* (*Cardiocarpus spinatus*). This further complicates the picture of evolution and introduction at this time. It is during the later phases of this brackish interval that tree ferns and medullosans became significant elements in lycopod-dominated swamps.

Second Dry Interval — Middle-Upper Pennsylvanian Transition

The consequences of the second dry interval are considerably different from those of the first because of the greater severity of climatic change. With the onset of second dry interval, the areal extent of coal swamps and their temporal duration diminished dramatically in comparison with those of the Middle Pennsylanian. There may even have been a brief period when coal swamps were almost nonexistent. We interpret the effects of the second, severely dry interval as follows. (1) Coal swamps were greatly reduced in areal extent and duration at least for some time prior to return of favorable conditions. (2) Swamps had much reduced areas of long-standing water. Seasonal moisture and runoff would have been sufficient to allow peat accumulation to occur, but parts of these swamps would probably have suffered long, seasonal dry-downs. (3) Vast areas would have been inundated by the sea during transgression, thus eliminating much of the potential swamplands.

The biological effects of the second dry interval are clear. All the major arborescent lycopods of the coal swamps and most of them from the clastic-swamp areas became extinct in the midcontinental coal basins. From this extinction we deduce that the wettest freshwater portions of coal swamps, which were dominated by arborescent lyco-pods, were eliminated. It is also probable that there were major extinctions of tree fern and medullosan species and that the effect of the climatic drying was the nearly complete elimination of coal swamps. The lack of detailed taxonomies for these latter groups makes interpretation of their patterns of change more difficult than for the lycopods. There was also a significant reduction in coal-swamp plant diversity and in the structural diversity of the vegetation. Tree ferns rose in importance and became dominant in most swamp communities, achieving the levels of dominance equal to or greater than those seen for lycopods in the late Middle Pennsylvanian. Medullosans were second in importance. The floristic similarity of the clastic lowland environments and coal swamps was high, and it is more difficult to differentiate coal-swamp from clastic-substrate vegetation after the second dry interval than at any other time during the Pennsylvanian. Marshes, dominated by *Chaloneria* with abundant *Sphenophyllum*, became ephemeral contributors to peat-forming wetlands.

The evolutionary consequences of the second dry interval are not clear. Our present taxonomic concepts for tree ferns and medullosans do not provide an adequate basis for determining their quantitative differences in late Middle Pennsylvanian and Upper Pennsylvanian coal swamps or differences between clastic and coal-swamp environments. However, the presence of these plant groups in Upper Pennsylvanian coal swamps may have been made possible because they were widely spread in many habitats of lowland regions. Thus complete extinction was unlikely. In general, the vegetation of Upper Pennsylvanian coal swamps was more heavily dominated by tree ferns than that of the surrounding mesic lowlands and clastic wetlands, where pteridosperms remained the dominant plants. The nearly complete elimination of lycopods may have occurred because they were ecologically centered in swamps and, with the nearly complete elimination of their adaptive zone, persistence was obviated. The only arborescent form to persist, *Sigillaria*, was centered in clastic-substrate areas. There were significant changes in the species composition of compression floras during the Upper Pennsylvanian as well (Gillespie and Pfefferkorn, 1979), implying that evolution was occurring in areas outside coal swamps. Considering the major extinctions of coal-swamp plants that had occurred, it is likely that the introduction of plants to the coal-swamp environments was common and may have changed markedly the species composition of all groups from that of the late Middle Pennsylvanian.

Cordaites and conifers began to appear in lowland assemblages in significant numbers during the Upper Pennsylvanain. They attained dominance only in some limnic coal swamps of western Europe (Galtier and Phillips, 1979). However, the presence of significant amounts of *Lycospora* in these limnic coal basins indicates that some arborescent lycopods continued to exist in Stephanian coal swamps, even after they became extinct in the midcontinent. Eventually, coal-swamp floras were composed mostly of tree ferns and pteridosperms and surrounded by a lowland vegetation dominated by coniferophytic gymnosperms, creating an extreme in terrestrial island refugia (Galtier and Phillips, 1979).

Conclusions

The in situ evolutionary component of coal-swamp vegetational change is very difficult to distinguish from the effects of introduction and/or increasing abundance of taxa that were centered in neighboring mesic lowland environments. Without a detailed understanding of the ecological distribution of a plant group during the Pennsylvanian Period, it will be difficult to discern the degree to which the coal-swamp record reflects the general direction of evolutionary change in that group. It appears that once a species was introduced and became abundant in the coal swamps it underwent little morphological change and gave rise to few other successful coal-swamp species. This supports the concept of coal swamps as "closed systems," environments that for much of the Pennsylvanian may have been evolutionary refugia. The low levels of evolutionary change that we detect, and the concentration of vegetational change to periods of climatic transition, suggest that the refugial nature of coal swamps was a consequence of the stressful nature of the abiotic environment, which supported few species relative to the general lowland region. This may have resulted in a dynamic equilibrium within a coal swamp, in which randomly appearing patches due to disturbance were most likely to be colonized from within the edaphically closed ecosystem. It was only during times of swamp-community disruption, due to isolation of swamps from each other during times of stress, that forms newly evolved within the closed system could survive or that introduced species could establish.

What we presently know of coal-swamp and clastic-substrate vegetation suggests that they were markedly different. We should not expect these different kinds of environments to provide us with the same kind of information about evolution or to be accurate reflections of the nature or absolute number of speciation events that occurred in the other. The climatic change during the early Middle Pennyslvanian is recorded more clearly by the floral composition of coal swamps than by that of the compression floras because the swamps were a more uniform and vegetationally isolated kind of habitat. In contrast, the Upper Pennsylvanian swamps recorded only a calamity and not the later subtle changes that ultimately altered the vegetation in the surrounding lowlands. Clearly, the best evidence of climatic and evolutionary changes that occurred during the Pennsylvanian is to be obtained from combined coal-ball, palynological, and compression studies in the context of paleoecological interpretations gleaned from geologic and biological evidence.

Summary

Major changes in the vegetation of Pennsylvanian coal swamps are related to changes in climate. Two episodes of reduced moisture availability, one during the early Middle Pennsylvanian and the other, more severe, during the early Upper Pennsylvanian, produced the most extreme changes in the coal-swamp flora: (1) expansion of indigenous swamp plants to dominant levels within the swamps; (2) introduction or migration of plants into coal swamps from surrounding lowlands; (3) extinctions; and (4) evolution of new species that subsequently became prominent in coal-swamp environments. Of these, the evolutionary changes are the most difficult to document. The edaphic properties of peat substrates tended to isolate coal-swamp plants from the surrounding clastic-substrate vegetation, making coal swamps edaphic islands within the terrestrial lowland setting. The evolutionary fates of lineages centered in coal swamps were largely independent of the changes occurring in environments surrounding coal swamps.

Five major groups of plants comprised coal-swamp vegetation: lycopods, cordaites, ferns, pteridosperms, and sphenopsids. The lycopods, which were most abundant during the Lower and Middle Pennsylvanian, probably contributed the largest number of endemic species. Cordaites were prominent only in somewhat drier or brackish coal swamps. Tree ferns and pteridosperms entered coal swamps during the first dry interval, expanded during the subsequent period of rising wetness — in the late Middle Pennsylvanian — then dominated the Upper Pennsylvanian swamps. Sphenopsids were of relatively minor importance in coal swamps. Evolution in all groups except the lycopods probably was centered in clastic-substrate lowlands or uplands; coal swamps preserve a fragmentary portion of the evolutionary history of the groups.

Acknowledgments

This study was supported in part by grants from the National Science Foundation (NSF–EAR–8018324) and the Donors of the Petroleum Research Fund (PRF–12412–G2) to W. A. DiMichele, by a grant from the NSF (NSF–EAR–7812954) to T.L. Phillips, and by the Illinois State Geological Survey. Figures were prepared by the Illinois State Geological Survey. We thank Suzanne Costanza, University of Illinois, for providing information on cordaitean organ assemblages, and Bruce Tiffney, Andrew Knoll, and Karl Niklas for comments on the manuscript.

References

Batenburg, L. H. 1981. Vegetative anatomy and ecology of *Sphenophyllum zwickaviense*, *S. emarginatum*, and other "compression species" of *Sphenophyllum*. *Rev. Palaeobot. Palynol.* 32:275–313.

———. 1982. "Compression species" and "petrifaction species" of *Sphenophyllum* compared. *Rev. Palaeobot. Palynol.* 36:335–59.

Baxter, R. W., and M. R. Willhite. 1969. The morphology and anatomy of *Alethopteris lesquereuxi* Wagner. *Univ. Kans. Sci. Bull.* 48:767–83.

Brack, S. D. 1970. On a new structurally preserved arborescent lycopsid fructification from the Lower Pennsylvanian of North America. *Am. J. Bot.* 57:317–30.

Broadhurst, F. M., I. M. Simpson, and P. G. Hardy. 1980. Seasonal sedimentation in the Upper Carboniferous of England. *J. Geol.* 88:639–51.

Chaloner, W. G. 1958. The Carboniferous upland flora. *Geol. Mag.* 95:261–62.

Conner, W. H., J. G. Gosselink, and R. T. Parrondo. 1981. Comparison of the vegetation of three Louisiana swamp sites with different flooding regimes. *Am. J. Bot.* 68:320–31.

Corsin, P. 1951. Basin houiller de la Sarre et de la Lorraine. I. Flore fossile, quatrième fascicule, Pecopteridees. *Études des Gîtes Mineraux de la France*, pp. 177–370 and pl. CVIII–CXCIX.

Costanza, S. 1981. A Middle Pennsylvanian cordaitean assemblage (*Cardiocarpus spinatus*) from the Illinois Basin [Abstr.]. *Bot. Soc. Am. Misc. Ser. Pub.* 160:43.

Cridland, A. A. 1964. *Amyelon* in American coal-balls. *Palaeontology* 7: 186–209.

DiMichele, W. A. 1978. Ecotypic variation in *Lepidodendron dicentricum* C. Felix [Abstr.]. *Bot. Soc. Am. Misc. Ser. Pub.* 156:32.

———. 1979a. Arborescent lycopods of Pennsylvanian age coals: *Lepidophloios*. *Palaeontographica Abt. B, Paläophytol.* 171:57–77.

———. 1979b. Arborescent lycopods of Pennsylvanian age coals: *Lepidodendron dicentricum* C. Felix. *Palaeontographica Abt. B, Paläophytol.* 171:122–36.

———. 1980. *Paralycopodites* Morey and Morey, from the Carboniferous of Euramerica — A reassessment of generic affinities and evolution of "*Lepidodendron*" *brevifolium* Williamson. *Am. J. Bot.* 67:1466–76.

———. 1981. Arborescent lycopods of Pennsylvanian age coals: *Lepidodendron*, with description of a new species. *Palaeontographica Abt. B, Paläophytol.* 175:85–125.

———. 1983. *Lepidodendron hickii* and generic delimitation in Carboniferous lepidodendrid lycopods. *Syst. Bot.* 8:317–33.

DiMichele, W. A., and T. L. Phillips. 1977. Monocyclic *Psaronius* from the Lower Pennsylvanian of the Illinois Basin. *Can. J. Bot.* 55:2514–24.

———. N.d. Arborescent lycopod reproduction and paleoecology in a coal-swamp environment of late Middle Pennsylvanian age (Herrin Coal, Illinois). *Rev. Palaeobot. Palynol.*, in press.

DiMichele, W. A., J. F. Mahaffy, and T. L. Phillips. 1979. Lycopods of Pennsylvanian age coals: *Polysporia*. *Can. J. Bot.* 57:1740–53.

Eggert, D. L., and T. L. Phillips. 1982. *Environments of plant deposition — Coal balls, cuticular shale, and gray-shale floras in Fountain and Parke Counties, Indiana*. Special Report 30. Bloomington: Indiana Geological Survey.

Ehrenfeld, J. G., and M. Gulick. 1981. Structure and dynamics of hardwood swamps in the New Jersey pine barrens: Contrasting patterns in trees and shrubs. *Am. J. Bot.* 68:471–81.

Felix, C. J. 1954. Some American arborescent lycopod fructifications. *Ann. Mo. Bot. Gard.* 41:351–94.

Galtier, J., and T. L. Phillips. 1979. Swamp vegetation from Grand 'Croix (Stephanian) and Autun (Autunian), France and comparisons with coal ball peats of the Illinois Basin. *Ninth International Congress of Carboniferous Stratigraphy and Geology, Urbana, Illinois, May 1979 — Abstracts of Papers*, p. 66.

Gillespie, W. H., and H. W. Pfefferkorn. 1979. Distribution of commonly occurring plant megafossils in the proposed Pennsylvanian System Stratotype. In *Proposed Pennsylvanian System Stratotype, Virginia and West Virginia*, ed. K. J. Englund, H. H. Arndt, and T. W. Henry, 87–94. Selected Guidebook Series no. 1. Falls Church, VA: American Geological Institute.

Johnson, P. R. 1979. Petrology and environments of deposition of the Herrin (No. 6) Coal Member, Carbondale Formation, at the Old Ben Coal Company Mine No. 24, Franklin County, Illinois. Master's thesis, Univ. of Illinois, Urbana-Champaign.

Koopmans, R. G. 1934. Researches on the flora of the coalballs from the "Aegir" Horizon in the Province of Limburg (The Netherlands). *Geologische Bureau voor het Nederlandsche Mijingebied te Heerlen, Jaarverslag over 1933*, pp. 45–46.

Leisman, G. A. 1960. The morphology and anatomy of *Callipteridium sullivanti. Am. J. Bot.* 47:281–87.

Mickle, J. E., and G. W. Rothwell. 1982. Permineralized *Alethopteris* from the Upper Pennsylvanian of Ohio and Illinois. *J. Paleontol.* 56:392–402.

Millay, M. A. 1979. Studies of Paleozoic Marattialeans: A monograph of the American species of *Scolecopteris. Palaeontographica Abt. B, Paläophytol.* 168:1–69.

———. 1982. Studies of Paleozoic Marattialeans: An evaluation of the genus *Cyathotrachus. Palaeontographica Abt. B, Paläophytol.* 180:65–81.

Oestry-Stidd, L. L. 1979. Anatomically preserved *Neuropteris rarinervis* from American coal balls. *J. Paleontol.* 53:37–43.

Parrish, J. T. 1982. Upwelling and petroleum source beds, with reference to the Paleozoic. *Am. Assoc. Pet. Geol. Bull.* 66:750–74.

Patterson, C. 1981. Significance of fossils in determining evolutionary relationships. *Annu. Rev. Ecol. Syst.* 12:195–224.

Peppers, R. A. 1970. *Correlation and palynology of coals in the Carbondale and Spoon Formations (Pennsylvanian) of the northeastern part of the Illinois Basin.* Illinois State Geological Survey Bulletin 93.

———. 1979. Development of coal-forming floras during the early part of the Pennsylvanian in the Illinois Basin. In *Depositional and structural history of the Pennsylvanian System of the Illinois Basin, Part 2 (Invited Papers)* ed. J. E. Palmer and R. R. Dutcher, 8–14. Guidebook to Fieldtrip 9, Ninth International Congress of Carboniferous Stratigraphy and Geology, Urbana, Il, May 1979. Urbana: Illinois State Geological Survey.

Peppers, R. A., and H. W. Pfefferkorn. 1970. A comparison of the floras of the Colchester (No. 2) Coal and Francis Creek Shale. In *Depositional environments in parts of the Carbondale Formation — Western and northern Illinois*, ed. W. H. Smith et al., 61–74. Guidebook Series, no. 8. Urbana: Illinois State Geological Survey.

Pfefferkorn, H. W. 1980. A note on the term "upland flora." *Rev. Palaeobot. and Palynol.* 30:157–58.

Pfefferkorn, H. W., and M. Thomson. 1982. Changes in dominance patterns in Upper Carboniferous plant-fossil assemblages. *Geology* 10:641–44.

Pfefferkorn, H. W., H. Mustafa, and H. Hass. 1975. Quantitative Charakterisierung ober-karboner Abdruckfloren. *Neues Jahrb. Geol. Paläontol., Abh.* 150:253–69.

Pfeiffer, R. N., and D. L. Dilcher. 1979. The paleobotany and paleoecology of the unnamed shale overlying the Danville Coal Member (VII) in Sullivan

County, Indiana. *Ninth International Congress of Carboniferous Stratigraphy and Geology, Urbana, Illinois, May 1979*—Abstracts of Papers, p. 163.

Phillips, T. L. 1979. Reproduction of heterosporous arborescent lycopods in the Mississippian-Pennsylvanian of Euramerica. *Rev. Palaeobot. Palynol.* 27:239–89.

———. 1980. Stratigraphy and geographic occurrences of permineralized coal-swamp plants — Upper Carboniferous of North America and Europe. In *Biostratigraphy of fossil plants*, ed. D. Dilcher and T. N. Taylor, 25–92. Stroudsburg, PA: Dowden, Hutchinson and Ross.

Phillips, T. L., and W. A. DiMichele. 1981. Paleoecology of Middle Pennsylvanian age coal swamps in southern Illinois — Herrin Coal Member at Sahara Mine No. 6. In *Paleobotany, paleoecology and evolution*, ed. K. J. Niklas, 1:231–84. New York: Praeger.

Phillips, T. L., A. B. Kunz, and D. L. Mickish. 1977. Paleobotany of permineralized peat (coal balls) from the Herrin (No. 6) Coal Member of the Illinois Basin. In *Interdisciplinary studies of peat and coal origins*, ed. P. N. Given and A. D. Cohen. *Geol. Soc. Am. Microform Pub.* 7:18–49.

Phillips, T. L., and R. A. Peppers. 1984. Changing patterns of Pennsylvanian coal-swamp vegetation and implications of climatic control on coal occurrence. *Int. J. Coal Geol.* 3:205–55.

Phillips, T. L., R. A. Peppers, M. J. Avcin, and P. F. Laughnan. 1974. Fossil plants and coal: Patterns of change in Pennsylvanian coal swamps of the Illinois Basin. *Science* 184:1367–69.

Phillips, T. L., J. L. Shepard, and P. J. DeMaris. 1980. Stratigraphic patterns of Pennsylvanian coal swamps in Midcontinent and Appalachian Regions of the United States. *Geol. Soc. Am. Abstr. Programs* 12:498–99.

Pickett, S. T. A. 1980. Non-equilibrium co-existence of plants. *Bull. Torrey Bot. Club* 107:238–48.

Raymond, A., and T. L. Phillips. 1983. Evidence for an Upper Carboniferous mangrove community. In *Second International Symposium on Biology and Management of Mangroves*, ed. H. Teas, 19–30. The Hague: Junk.

Reihman, M. A., and J. T. Schabilion. 1979. Petrified *Neuropteris scheuchzeri* pinnules from the Middle Pennsylvanian of Iowa: Paleoecological interpretation. *Ninth International Congress of Carboniferous Stratigraphy and Geology, Urbana, Illinois, May 1979 — Abstracts of Papers*, p. 177.

Rothwell, G. W., and S. Warner. N.d. *Cordaixylon dumasum* n. sp. (Cordaitales): I. Vegetative structures. *Bot. Gaz.*, in press.

Schlesinger, W. H. 1978. Community structure, dynamics and nutrient cycling in the Okefenokee cypress swamp-forest. *Ecol. Monogr.* 48:43–65.

Schopf, J. M. 1973. Coal, climate and global tectonics. In *Implications of continental drift to the earth sciences*, ed. D. H. Tarling and S. K. Runcorn, 1:609–22. New York: Academic Press.

———. 1975. Pennsylvanian climate in the United States. In *Paleotectonic investigations of the Pennsylvanian system in the United States — Part 2: Interpretive summary and special features of the Pennsylvanian System*, ed. E. D. McKee and E. J. Crosby. *U.S. Geol. Surv. Prof. Pap.* 853:23–31.

————. 1979. Paleoclimatology and paleontology. In *Depositional and structural history of the Pennsylvanian System of the Illinois Basin, Part 2 (Invited Papers)*, ed. J. E. Palmer and R. R. Dutcher, 1. Guidebook to Fieldtrip 9, Ninth International Congress of Carboniferous Stratigraphy and Geology, Urbana, Illinois, May 1979. Urbana: Illinois State Geological Survey.

Scott, A. C. 1977. A review of the ecology of Upper Carboniferous plant assemblages, with new data from Strathclyde. *Palaeontology* 20:447–73.

————. 1978. Sedimentological and ecological control of Westphalian B plant assemblages from West Yorkshire. *Proc. Yorks. Geol. Soc.* 41:461–508.

Shadle, G., and B. M. Stidd. 1975. The frond of *Heterangium*. *Am. J. Bot.* 62:67–75.

Snigirevskaya, N. S. 1972. Studies of coal balls of the Donets Basin. *Rev. Palaeobot. Palynol.* 14:197–204.

Stidd, B. M., L. L. Oestry, and T. L. Phillips. 1975. On the frond of *Sutcliffia insignis* var. *tuberculata*. *Rev. Palaeobot. Palynol.* 20:55–66.

Stidd, B. M., and T. L. Phillips. 1973. The vegetative anatomy of *Schopfiastrum decussatum* from the Middle Pennsylvanian of the Illinois Basin. *Am. J. Bot.* 60:463–74.

Thomas, B. A. 1967. *Ulodendron*: Lindley and Hutton and its cuticle. *Ann. Bot.* n.s. 31:775–82.

————. 1970. Epidermal studies in the interpretation of *Lepidodendron* species. *Palaeontology* 13:145-73.

Wanless, H. R., J. B. Tubb, Jr., D. E. Gednetz, and J. L. Weiner. 1963. Mapping sedimentary environments of Pennsylvanian cycles. *Bull. Geol. Soc. Am.* 74:437–86.

Wanless, H. R., J. R. Baroffio, and P. C. Trescott. 1969. Conditions of deposition of Pennsylvanian coal beds. In *Environments of coal deposition*, ed. E. C. Dapples and M. E. Hopkins. *Geol. Soc. Am. Spec. Pap.* 114:105–42.

Wartmann, R. 1969. Studie über die papillen-förmigen Verdickungen auf der Kutikule bei *Cordaites* an Material aus den Westfal C des Saar-Karbons. *Argumenta Palaeobotanica* 3:199–207.

Williamson, P. G. 1981. Palaeontological documentation of speciation in Cenozoic molluscs from Turkana Basin. *Nature* 293:437–43.

Winston, R. B. 1983. A late Pennsylvanian upland flora from Kansas: Systematics and environmental implications. *Rev. Palaeobot. Palynol.* 40:5–31.

Zangerl, R., and E. S. Richardson, Jr. 1963. *The paleoecological history of two Pennsylvanian black shales*. Fieldiana, Geology Memoirs 4. Chicago: Chicago Natural History Museum.

Zeigler, A. M., R. K. Bambach, J. T. Parrish, S. F. Barrett, E. H. Gierlowski, W. C. Parker, A. Raymond, and J. J. Sepkoski, Jr. 1981. Paleozoic biogeography and climatology. In *Paleobotany, paleoecology and evolution*, ed. K. J. Niklas, 2:231–66. New York: Praeger.

MICHAEL J. COPE AND
WILLIAM G. CHALONER

Wildfire: An Interaction of Biological and Physical Processes

Introduction

The occurrence of forest fire (or, more correctly, wildfire in general) has the distinction of being a purely physical process that is nonetheless totally dependent on biological productivity. In this, it is a phenomenon for which we have abundant evidence in geological history and which records a range of interactions between physical and biological processes. The occurrence of wildfire has four distinct but interrelated requirements: (1) the accumulation of fuel in the form of plant material, produced by photosynthetic carbon fixation; (2) a climate favorable to plant material reaching a dry enough state, at least seasonally, to ignite; (3) an atmospheric composition in which combustion can occur; and (4) some mechanism of ignition. All four must be fulfilled for wildfire to take place. Evidence from a variety of sources suggests that wildfire has been a feature of the natural environment for at least 350 million years, almost as long as plants have existed on land (Cope, 1981; Kemp, 1981; Komarek, 1972; Scott, 1980). This article reviews some of the implications of this suggestion, in terms both of geological (physical) phenomena and those of a purely biological nature.

Characteristics of Charcoal — The Solid Residue of Wildfire

Wildfire, Combustion, and Pyrolysis

The most clear-cut and paleobotanically informative product of wildfire is charcoal (structured fusinite of petrographical usage). Charcoal is produced when plant material is raised to a temperature at which its more volatile constituents are driven off (pyrolysis), leaving a chemically altered residue of high carbon content. Laboratory studies have indicated that the thermal degradation of woody (ligno-cellulosic) tissue may be categorized into a number of discrete temperature zones characterized by the generation and expulsion of different gaseous and solid products. For a detailed review of this work, see Beall and Eickner (1970).

At temperatures up to 200°C, noncombustible products are evolved — water (completely removed at 140°C), carbon dioxide, formic acid, acetic acid, and glyoxal. Combustible carbon monoxide is first produced at 200°C. At 225°C flame point is reached and any combustible gases will burn if an ignition source is present. From 280°C to 500°C active pyrolysis takes place with the generation of the combustible gases carbon monoxide, methane, and hydrogen; also in this temperature range highly flammable tars are produced. Above 330°C the flash point of the combustible gases is passed and, in the presence of adequate oxygen, these gases are then subject to spontaneous ignition. As combustible gases and tars are expelled, charcoal forms as a solid residue. Above 500°C charcoal will become increasingly carbon-rich with rising temperature, as the atomic oxygen and hydrogen of the original ligno-cellulosic material becomes steadily depleted. Above 1,000°C only carbon monoxide and hydrogen are produced.

Under wildfire conditions in the presence of unrestricted oxygen supply, combustion results in the total destruction of the plant material, leaving only mineral ash as a residue. Charcoal formation will occur in cases of more restricted oxygen supply where burning is impeded, as for example in soil litter beneath burning plant debris or in the inner parts of large tree trunks, of which the outer parts may have burnt freely. The burning process, which causes the necessary rise in temperature, generally destroys far more material than is preserved as charcoal. Nonetheless, the residual charred product is important as tangible evidence of the pyrolysis process and as a unique and informative type of preservation of plant material. Charring of plant tissue evidently makes it less "biodegradable" than the original, more labile state, and this resistance to microbial attack increases its prospects of survival as a fossil. Small charred fragments are also capable of considerable transport and retain their internal structure to a high degree on subsequent burial.

Macroscopic Features of Fossil Charcoal

Structural features of fossil charcoal, visible on inspection of the material by eye, form a basis for discriminating between black, coalified plant tissue (vitrinite *sensu lato*) and charcoal. Harris (1981) cites five features that distinguish Mesozoic fossil leaves preserved as charcoal from those preserved as (coalified) compression fossils.

1.   The charred leaves seldom exceed 2 cm and have sharply truncated broken ends; coalified leaves may be far longer, may have been bent in the burial process, and sometimes have frayed ends.

2.   Fragments of the charred leaves are black and totally opaque, whereas those of coalified leaves are browner in color and, if thin enough, translucent.

3.   Broken surfaces of charred leaves show cellular detail, often exposed by a fracture passing through the leaf tissue. Leaf compressions are normally of homogeneous (noncellular) character on fracture and are typically exposed by a fracture plane passing between leaf and matrix rather than through the coaly material itself.

4.   Charcoalified plant material is very resistant to oxidative maceration, although after several weeks it may succumb. Coalified material oxidizes more readily, and subsequent alkali treatment may free intact cuticles, if present.

5.   Charred fossil material has little volatile content and glows on combustion, whereas vitrinitized plant tissue burns with a smoky flame.

To these features, evident in leaf material, may be added the macroscopic features peculiar to wood charcoal. As Harris (1958) demonstrates, the combination of shrinkage cracks and brittleness following charring causes twigs to break up into short segments. Uncharred twigs of the same diameter do not break up so readily during water transport. Harris showed that charred twigs from a natural (wildfire) setting showed a mean length-to-breadth ratio of 1.6, compared with 33 for uncharred twigs. Similarly, larger pieces of charred wood show strong transverse and radial fractures that characteristically produce more or less cuboidal blocks on breaking up (figs. 9.1A and 9.1B).

Microscopic Features of Fossil Charcoal

On a microscopic scale, charred plant tissue may retain a high degree of three-dimensional structure, with cell lumina open even when charred fragments occur in a matrix of coaly (vitrinitized) material, as in a bituminous coal. Closure of the cell lumina due to compression of charcoal occurs only with the shattering of cell walls to produce a characteristic "bogen structure" (from the German *Bogen*, a bow or curve). That is, the shattered woody tissue shows in transverse section a series of packed, interlocking curved cell-wall fragments, produced by

Fig. 9.1. A. Piece of charred *Pinus* branch from wildfire, Dumfriesshire, Scotland, showing typical shrinkage cracks resulting from charring, which produce more or less cuboidal fragments (actual size). B. Fragment detached from the same specimen, showing sharp transverse fractures and wood grain. (4X). C. Fragment of *Callixylon* fusinized wood, Berea Sandstone, Lower Mississippian, near Amherst, Ohio (from Professor Charles Beck), showing similar grain and blocky character but with some rounding, probably caused by transport (4X).

brittle fracture under load. This contrasts with the appearance of coalified, compressed wood (vitrinite), in which the cell cavities have become closed without their walls shattering.

The opaque nature of fossil charcoal presented great difficulties to early microscopists, both paleobotanists and coal petrographers. Moreover, its inert character meant that even lengthy oxidative maceration generally failed to render it translucent for investigation. However, the application of scanning electron microscopy (SEM) has effected an important breakthrough in this field (Muir, 1970; Scott and Collinson, 1978). A number of SEM studies of charcoalified fossils have illustrated the preservation of small plant organs yielding details of internal structure. Mesozoic ferns (Alvin, 1974), a Westphalian conifer (Scott, 1974; Scott and Chaloner, 1983), and Cretaceous flower buds (Friis and Skarby, 1981, 1982) are examples of plant structures preserved by the charring process that have yielded anatomical detail under SEM. Alvin's material showed such features as the three-dimensional nature of the leaf mesophyll and the stomatal guard cells and subsidiary cells. Such information could otherwise be obtained only from fossil leaf material in a permineralized state of exceptionally high-quality preservation. The flowers described by Friss and Skarby are equally remarkable in the detail they show, including such delicate structures as stamens, stigma, and even nectaries. These Cretaceous fossils preserved in charcoal are the oldest flowers for which we have such complete and detailed knowledge.

Evidence of a Pyrolytic Origin for Charcoalified Plant Fossils

Detailed microscopic studies using SEM have provided a basis for the recognition of pyrolyzed plant tissues in both the archeological and the geological record. In the last decade a considerable amount of information has been generated by the work of wood and fuel technologists concerned with industrial charcoal manufacture. It is from this literature and its application to the fossil record that we have derived our current views as to the pyrolytic origin of fossil charcoal.

From an SEM study of charcoals produced at 270–400°C under industrial kiln conditions, McGinnes et al. (1971, 1974) noted the following structural features of charcoalified wood: (1) development of radial splits on a macroscopic scale; (2) shrinkage in the tangential, radial, and longitudinal planes; (3) the presence of residual tars in cell cavities; (4) destruction of the original microfibrillar arrangement of the wood of produce a smooth, amorphous inner cell wall, which may display marked folds; and (5) apparent cell-wall homogenization with

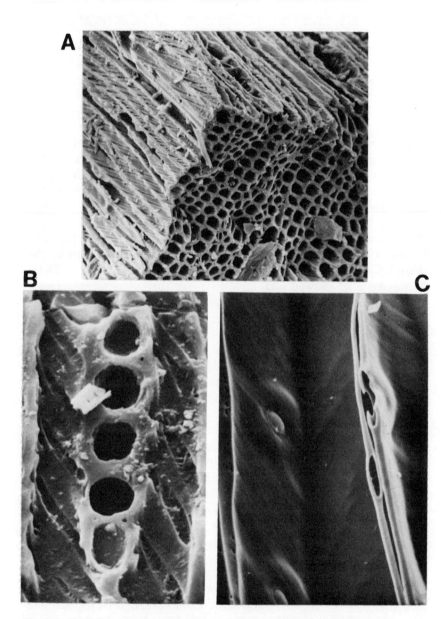

Fig. 9.2. Scanning electron micrographs of fossil and modern conifer charcoal. A. Transverse (lower right) and radial longitudinal (left) faces of charcoal from a Rhaeto-Liassic fissure deposit at Cnap Twt, Glamorgan, South Wales, showing pronounced checking of the tracheid walls on the radial face (400X). This wood is probably of the same type as that attributed by Harris (1957) to the conifer *Hirmerella*

elimination of the middle lamellar zone, at least as a feature recognizable under SEM.

There is good evidence in fossil charcoal specimens for the presence of fine structural features seen in the synthetic high-temperature chars described by McGinnes and co-workers.

Radial-split development and shrinkage are shown by many charcoalified fossil wood specimens. Shrinkage along the longitudinal axis accounts for the more or less cuboidal nature of many larger charcoal fragments (see fig. 9.1). The development of radial splits can be inferred from the general platelike morphology of smaller charcoal fragments, which reveal the traces of wood rays on their two largest surfaces. Real dimensional change has been documented by Alvin, Fraser, and Spicer (1981), who demonstrate shrinkage in fusinitized specimens of *Protopodocarpoxylon* wood, compared with mineralized specimens of the same genus.

The presence of residual tars has not been observed in fusinite specimens examined by the authors. However, this feature may well be the cause of the occurrence, puzzling to coal petrographers, of the association of resinous bodies with fusinities (White and Thiessen, 1913; Schopf, 1948, 1975).

Cope (1980) and Cope and Chaloner (1980) use the well-defined microscopic feature of cell-wall homogenization seen under SEM as a criterion for the recognition of charcoal in the fossil record. Although in general a smooth, amorphous inner face to the cell wall may also be seen in fossil charcoals, helical striations corresponding to microfibrillar orientation may survive charring (see fig. 9.2C).

It is appropriate to consider here the range of temperature conditions that may develop during the occurrence of wildfires. Data from a number of sources (Beall and Eickner, 1970; Brown and Davis, 1973; Hare, 1961) indicate that a wide range of temperatures is achieved, according to whether the temperature is sampled at the ground surface, above rising smoke columns, or within tree trunks. Clearly, wood subjected to elevated temperatures during wildfire will show a variety of responses; some will combust completely, some will be charcoalified

---

muensteri. (Photo reproduced by kind permission of Mrs. Ruth Bennett, University of Bristol.) B. Charcoalified wood, middle Jurassic of the Scalby Formation, Yorkshire, showing a tangential face with a ray (center) and checking cracks in the adjoining tracheids (1,000X). C. *Pinus sylvestris* after charring in a muffle furnace at 500°C for one hour (2,000X). Note the obliquely striated wall surface reflecting the microfibrillar structure.

under high-temperature pyrolysis, and some will be quickly scorched at lower temperatures. Thus the products of wildfire (charcoals) will show more variation in their structural preservation than would synthetic charcoals prepared under constant temperature conditions for long "soak" times (that is, prolonged exposure to charring temperature). We would expect that scorched material would not show obliteration of the middle lamella and that microfibrillar detail would be preserved; the transition to true charcoal would result in the elimination of both these features of structural detail (which is the state seen in the majority of fossil charcoal specimens). It is interesting to note that remnant traces of a middle lamella are to be seen in the semifusinite portions of vitrinite-fusinite "transitional" material that occurs in bituminous coals; such occurrences are consistent with an origin as scorched material.

Fusinite Attributed to Biological Degradation

A controversy, extending back some seventy years, has centered around the possibility that structured fusinite might have been formed in the past by some biological process involving oxidative degradation without combustion or other pyrolysis being involved. Schopf (1975, 45) sets out the case against "the charcoal theory of the origin of all fusain." With characteristic candor, however, he admits the inherent weakness in any argument for a nonfire origin of structured fusinite in saying "it is unfortunate that I am not able to suggest any generally applicable alternative explanation."

Beck, Coy, and Schmid (1982) have made a recent contribution to the nonfire-origin case in their account of fusinitized material of the progymnosperm wood *Callixylon* from the Mississippian of Ohio. They state that fusain in the fossil record "probably had several origins, including forest fire." They further suggest that (potential) "natural sites of fusain production in the absence of oxygen are regions of vulcanism and organic sediments inhabited by anaerobic microorganisms" (p. 54). It is in this latter proposition that we would differ from those authors. In the entire debate on the nonfire origin of fusinite, none of the proponents for that thesis has ever demonstrated the formation of fusinitized wood showing the features enumerated by McGinnes et al. (1971, 1974; see description above) by microbial degradation occurring in a modern sedimentary environment. Indeed, Cohen and Spackman (1977, 72). after a study of some modern swamp-peat environments, state: "No evidence is found to support the hypothesis that fusinite in coal can be derived from any process other than fire."

Beck, Coy, and Schmid draw particular attention to two features of their fusinitized *Callixylon* that they suggest favor a nonpyrolytic origin. As evidence they point, first, to the microfibrillar structure on the cell-wall surfaces and, second, to features they describe as "checking." Checking cracks may be broadly defined as fissures in the walls of wood elements that generally parallel the helical configuration of cellulose microfibrils forming the wall. Indeed, the occurrence of checking is the result of structural weakness of the wall between the constituent microfibrils. Such checking cracks may occur in modern wood on drying out, particularly if some degradation of the wall material has occurred.

Professor Charles Beck has very generously made a sample of the fusinitized *Callixylon* wood available to us for examination, and we confirm their observations. A fragment of their material, showing a characteristic cuboidal form and evident woody (cellular) structure, is shown in figure 9.1C. We do not regard the evidence for microfibrillar structure in the wall as being in itself evidence for a nonpyrolytic origin. *Pinus* wood subjected to pyrolysis in a muffle furnace at 500°C for an hour shows clear features of microfibrillar wall striations (fig. 9.2C), although it was rendered black and showed other typical charcoal characteristics.

Checking, involving the opening up of cracks in the tracheidal walls, is evidently not obliterated by high temperature; once cracks have opened up in wood elements, nothing short of fusion of the cell-wall material could eliminate them. There are certainly longitudinal cracks in the tracheids of *Callixylon* fusinite from Ohio, presumably induced by shrinkage, which may have occurred either before or during the pyrolysis process. Checking cracks are common in fossil fusinites showing other characteristic features of pyrolyzed wood. The fossil charcoal attributed by Harris (1957) to the conifer *Hirmerella* from fissure deposits in South Wales, from which he adduced evidence for Mesozoic forest fire, shows typically unflattened structure and prominent checking cracks (see fig. 9.2A). Jurassic fusinite showing homogenized cell walls and open cell lumina also show clear checking cracks (fig. 9.2B).

In summary, we feel that the fusinitized *Callixylon* wood examined by Beck, Coy, and Schmid (1982) shows features that are in all respects consistent with charring produced by wildfire conditions. The specimen they describe was probably pyrolized at about 250°C (significantly above the normal range of diagenetic temperatures), a temperature sufficient to render the specimen inert to microbial degradation but not high enough to eliminate the microfibrillar topography of the inner wall surfaces. We can see no reason to abandon our belief that structured fusinite is a product of pyrolysis, representing most commonly the product of wildfire.

Charcoal Produced by Extrusive Volcanic Activity

Clark and Russell (1981) point out that occurrence of charcoal in the fossil record is not a direct proof that combustion has occurred, since engulfing of trees by a lava flow, or possibly by other hot, extrusive volcanic products, may pyrolyze plant material without combustion occurring. However, such occurrences seem to have been relatively rare as a source of charcoal in the fossil record. Indeed, trees, engulfed in lava flows are usually completely destroyed and a lava-filled tree mold occupies the resultant void (for example, Lockwood and Williams, 1978). Where charcoal occurs abundantly in a clastic or carbonaceous sediment lacking obvious volcanic or pyroclastic constituents, it is a reasonable assumption that it was a product of wildfire rather than a volcanic process. As Clark and Russell remark, "in the remote past, heat from wood fires, sustained by atmospheric oxygen, was probably the dominant agent for a charcoal production" (p. 428).

Sources of Ignition of Wildfire

The principal natural source of ignition of wildfires (that is, discounting human intervention) is evidently lightning strike. However, other sources that have been documented include volcanic activity, meteorite fall, spontaneous combustion of decaying organic material, and sparks caused by rock falls.

Possible mechanisms for the natural ignition of wildfire in the pre-Quaternary are reviewed by Kemp (1981) and touched on briefly by Cope and Chaloner (1980). Kemp reports that 80 percent of wildfires in semiarid western Queensland have been attributed to lightning. The involvement of people in igniting forest fires at the present day has perhaps distorted our picture of the potential of ignition from natural sources. The Quaternary evidence from lake-sediment sequences shows that the role of early human activity may have been very different in this respect in different climatic regimes. Evidence from lake sediments in Ontario, Canada, shows that, during the last 225 years, the frequency of forest fire has increased from an average of one major fire every 80 years to one every 14 years, and this is presumably to be attributed to human intervention (see Moore, 1982). Singh, Kershaw, and Clark (1981) compare pollen and charcoal abundance in lake sediment in the more arid environment of the central Australian desert. They suggest that in that setting the human role in starting wildfire was insignificant, "as there may have been little scope for increasing (natural) fire frequencies in the arid zone" (p. 48). Terasmae and Weeks (1979, 116), in a study of charcoal in Canadian lake sediments, relate incidence of detrital charcoal and pollen spectra and conclude that "a climatic control of forest fires is (to be) inferred." A helpful review of the growing literature on the Quaternary record of forest fire is given in Terasmae and Weeks (1979) and in Singh, Kershaw, and Clark (1981). There seems little doubt that climate and its control over the climax vegetation has been the overriding factor governing the incidence of wildfire in time past and that in present-day environments, from temperate to tropical latitudes, lightning strike is far more significant as a natural source of wildfire ignition than all other natural sources put together.

The Occurrence of Wildfire through Geological Time

As soon as plant material was being formed on land by a terrestrially adapted flora, burning of such material would have been possible, given suitable conditions. Accumulations of algal debris on strandlines might even, on drying out, have presented an ignitable fuel source, long before the Silurian colonization of the land by vascular plants. At some stage, perhaps in the Lower Paleozoic, or maybe not until Devonian time, there must have been a first occurrence of wildfire involving the burning of plant material. Although such burning might not necessarily have resulted in a charred residue, we would expect that the first charcoal production and the earliest wildfire would be closely linked in time. Edwards (1982, fig. 60) has illustrated early Devonian specimens of the putative land plant *Prototaxites*, showing open cell lumina and apparently homogenized cell walls. It is obviously hazardous to argue by analogy with charcoalified modern woods that these specimens represent charred material, since we know so little of the original nature of this plant. Although it is tempting to see such material as evidence of Devonian wildfire, we certainly need a less ambiguous basis for such a first record. At present, the earliest well-authenticated occurrences of charcoal appear to be Lower Carboniferous (Beck, Coy, and Schmid, 1982; Scott and Collinson, 1978).

Charcoal (as structured fusinite) is certainly a common constituent of bituminous coals from the Carboniferous onwards. Teichmuller (1975, 219) states "most fusinites of German and Australian soft brown coals are pyrofusinites. In the Euro-American Carboniferous coals the macroscopically visible fusain lenses and layers result from fires." Charcoal also occurs in the Permo-Carboniferous coals of Gondwanaland (Falcon, 1978). Our own observations of Carboniferous sediments, other than coal-bearing sequences, demonstrate the widespread occurrence of dispersed charcoal in a variety of facies, including shelf limestones and fluvio-deltaic clastics. Detrital wood charcoal is recorded as a significant constituent of Triassic (Harris, 1957, 1958), Jurassic (Cope, 1981), and Cretaceous (Alvin, 1974; Alvin, Fraser, and Spicer, 1981; Friis and Skarby, 1981, 1982; Harris, 1981) rocks where volcanic or pyroclastic pyrolysis seems improbable from the nature of the associated sediments. Grebe (1953) has documented the changes in contemporaneous vegetation shown by the palynological record associated with charcoal-rich layers in Tertiary brown coals. Kemp (1981) gives a very full review of evidence for the occurrence of pre-Quaternary wildfire in Australia. However, although she acknowledges the occurrences of "fusain" in Australian coals ranging in age from Permian to Tertiary, she is guarded in attributing a pyrolytic origin to them all. It is interesting to note in passing that Quaternary workers are generally more unreserved in attributing a fire origin to any fusinite/charcoal encountered in rocks of that period.

It would be of the greatest interest to obtain quantitative data indicating whether the frequency of incidence of wildfire has changed significantly through time, against the background of well-documented changes in climate and land vegetation. Quantitative assessments have been made in modern Quaternary lake sequences where the nature of the catchment and associated vegetation are relatively well understood (for example, Singh, Kershaw, and Clark, 1981). In order sediments, the variables (such as differential destruction of charred versus uncharred plant debris) may completely mask original differences associated with a changing fire regime. Although it is difficult to assess the distribution of charcoal in the fossil record in a quantitative manner, it is clear that some sedimentary sequences contain abnormally high amounts of charcoal. Moreover, in many such sequences it appears that charcoal occurrence, compared with the flora seen in compression fossils, is to some degree cyclic. This may be a manifestation of sedimentological processes, or it may be related to true periodicity in charcoal production and therefore in wildfire frequency. Good evidence of such cyclicity is seen in the distribution of "fire horizons" (fusain layers) in the Tertiary brown coals of northern Germany (Teichmuller, 1975). Similar patterns of cyclic increase in the fusain content of coal seams are also noted in Paleozoic bituminous coals (Schopf, 1952). However, it must be borne in mind that in many instances where discrete charcoal horizons are identified, the sediments concerned generally accumulated in continental to marginal marine environments.

These kinds of environments are subject to rapid changes in sedimentary conditions, which have the capacity to obscure real cyclic events. Under open marine conditions sedimentary facies are relatively stable over long periods of time and offer a better means of assessing periodicity in the occurrence of particular phytoclast types. Summerhayes (1981) has studied organic facies in Jurassic-Cretaceous sediments of North Atlantic deep basins. He records that "fusain" (charcoal) shows periodic increases in abundance relative to other organic constituents over long intervals of geological time. Although he attributes this situation to the removal (by oxidation) of the more labile constituents of the organic assemblage, his data provide evidence that charcoal occurs generally as 1–10 percent of the total organic fraction in Jurassic-Cretaceous oceanic sediments. It is our interpretation of these data that wildfire was indeed a regular and frequent occurrence in Mesozoic times. In fact, although the documentation of structured fusinite through the geological column is inevitably still very incomplete, the available evidence suggests that wildfire has been a feature of the terrestrial environment from at least the Carboniferous until the present day, without any evident interruption.

Implications of Wildfire

Paleoatmospheric Composition

The occurrence of fossil charcoal has important potential as an indication of the existence of an atmosphere capable of sustaining combustion. If, as argued above, structured fusinite is seen as a product of pyrolysis, its existence required temperatures high enough to bring this about. In the absence of indicators of a volcanic (igneous) source of heat, wildfire is the only possible causal agent for such elevated temperature (Cope and Chaloner, 1980). Wildfire requires a minimal concentration of oxygen below which combustion cannot occur. The combustion of wood involves the burning of carbon monoxide and methane, generated by the heat produced in the combustion process, in an atmosphere containing oxygen. On the basis of work by Coward and Jones (1952) on the burning of methane and carbon monoxide in oxygen/nitrogen (air) mixtures, the present authors argued (1980) that combustion involving methane and carbon monoxide would not occur in an atmosphere of less than 0.3 of the present atmospheric oxygen level (PAL). However, K. N. Palmer has since pointed out (pers. com., cited in Cope and Chaloner, 1981) that Coward and Jones's data were based on a study of combustion in which the gases involved were premixed. A more realistic basis for modeling the conditions of wildfire would be one in which the fuel gases meet the oxygen atmosphere without premixing. In these circumstances the minimal level of oxygen required to sustain combustion is probably nearer 13 percent (that is, 0.6 PAL). This figure is considerably higher than some estimates of oxygen levels in the Phanerozoic atmosphere. Tappan (1968), for example, has suggested on the basis of fluctuations in planktonic oxygen production by photosynthesis that the atmospheric oxygen may have fallen well below this level during the Phanerozoic, and notably during the Permian to Jurassic interval. The abundant occurrence of fusinite in Gondwana (Permian) coals and the frequent presence of charcoal fragments in Jurassic sediments is in conflict with this hypothesis.

Obviously, if there have been significant changes in the oxygen content of the atmosphere, they would have had important physiological consequences in terms of plant and animal respiration. It is not our intention to explore these consequences here, but it is worth noting that the atmospheric composition required for combustion involves very different considerations from those which affect the gas exchange in animal respiration. Clark and Russell (1981) point out that the critical threshold for burning is not simply the partial pressure of oxygen (which governs its availability for gas exchange) but the absolute proportion of oxygen to inert gas. They claim that if inert gas is added to an atmosphere with an adequate partial pressure for human respiration, the absolute percentage of oxygen may be "reduced sufficiently to prevent a match from burning." Most discussions of the paleo-atmospheric composition seem to assume a more or less constant value for the earth's atmospheric pressure. If this has changed drastically through the course of Phanerozoic time, a wide range of meteorological, weathering, and other phenomena would also have been affected. We know of no suggestion that this has occurred. It can only be said that unless there have been very considerable changes in the total atmospheric pressure, which on available evidence seems unlikely, there has been an oxygen content of the atmosphere of not less than 60 percent of its present level since the late Devonian/early Carboniferous.

Geological Processes

Two obvious and immediate effects of a geological nature must have accompanied forest fire in the past. These are the change of runoff under a given rainfall regime and the increase in erosion rate, both resulting from sudden removal of vegetational cover. It is difficult to assess the magnitude of these processes in the geological past, but work on modern environments makes it clear that these effects must have been considerable. Moore (1978) reports an increase in runoff of 60 percent in a Minnesota lake following forest fire. This was apparently caused by loss of the plant cover and its associated functions of transpiration and interception of rainfall. The increased erosion resulting from forest fire can be even more dramatic. Swanson (1981) has shown that a thirtyfold increase of soil erosion may follow the occurrence of a forest fire. Although it is possible — indeed likely — that such surges of erosion may have followed extensive forest fire in the past, we know of no documentation linking such occurrences to fire in any sedimentary sequence older than the Quaternary. This may of course be a result of an understandable tendency among geologists to attribute any perceptible alterations in sedimentation regime to a change in base level or to physical phenomena other than forest fire. An investigation of sedimentation phenomena, detrital charcoal, and palynological evidence from pre-Quaternary sequences might well elicit earlier evidence of this link between fire and erosion (see, for example, Grebe, 1953, whose work is considered above).

Biological Consequences

The ecological consequences of wildfire have received much attention (for recent reviews see Moore, 1982; Gill, Groves, and Noble, 1981; Rundle, 1981). In considering the biological role of forest fire in the past, it is helpful to separate immediate effects (removal of plant cover, opening up the habitat to colonization, release of nutrients) from the long-term ones that constitute selective pressures favoring the evolution of structures and strategies that facilitated survival under a fire regime.

A very readable review of what constitutes such adaptation to fire is given by Bradstock (1981). In some well-documented instances, the adaptation of certain plants to fire survival is very evident. In the so-called closed-cone pines of North America, for example, the cones do not open to release their seeds until they have been scorched by fire (Raven, Evert, and Curtis, 1981). A similar fire control of seed release is witnessed in Australian *Banksia* species. A comparable result may be achieved by those fire-tolerant species of which the seeds, lying in the soil, have their dormancy broken by their exposure to fire (Bradstock, 1981).

Raup (1981, 39) argues that throughout time disturbance of plant communities by various physical forces (of which fire is only one) has been an important element in the evolution of plant populations. Discussing the role of fire, he suggests that "wherever the forests are of resinous needle-leaved trees the principal disturbing agent is fire started by lightning or people." The same can surely be said of the very different sclerophyll forests of Australia (Ashton, 1981, 362). But as that author points out, "the great flammability of the forest may be the secret of the success of the forest regeneration.... In one sense the tolerance of the high forests to severe fire once every 1, 2 or 3 centuries is due to their low resistance to it. It is thus somewhat analogous to the strategies of some plants in relation to obligate pathogens where the most tolerant hosts possess the most hypersensitive tissues." A some-what different analogy is perhaps to be seen in the grass/grazer relationship. The grass sustains the grazer, which damages the grass by partially defoliating it, but grazing removes grass competitors (for examples, tree seedlings) in the process. Leaf regeneration by growth from the base enables the grass to survive this predation. The subclimax grassland is thus sustained by a highly destructive but selective biotic factor, the grazer, which is in turn sustained by the grass it consumes. The relationship of sclerophyll forest to fire thus parallels that of grassland to grazing fauna.

The fire-proneness of certain types of trees and their foliage raises the question of how such seemingly vulnerable attributes ever evolved. First, it needs to be acknowledged that certain secondary products of plants may have enormous adaptive advantage at one phase in the plant's life and be disadvantageous at another. It is likely that the high terpenoid content of conifer leaves was selected for, in the first instance, a protection against microbial attack or animal (insect or browser) predation, or perhaps both (Swain and Cooper-Driver, 1981). The advantage of this protection may have far outweighed the disadvantage of resulting increased vulnerability to fire. Even this disadvantage, however, is perhaps illusory. If high flammability is coupled with a regeneration strategy that can cope with fire (for example, fruiting structures that are fire-resistant or, indeed, release seed only after fire exposure), then adaptations enhancing flammability themselves become advantageous. For as noted above, fire can remove competitors that, without its intervention, might have superseded the more fire-prone (but fire-adapted) forest. Evidence for a rather finely tuned example of such fire adaptation is cited by Williamson and Black (1981). It has long been noted that in many temperate ecosystems the normal succession from pine forest to hardwood forest may be interrupted by fire to produce a so-called fire climax of conifer woodland. The critical control of which forest constituent survives operates when the pine and hardwood element are competing at a mixed oak-pine stage. Williamson and Black show that maximum temperatures recorded during fires in this mixed woodland "were sufficiently higher under pines than under oaks to ensure elimination of the competitively superior oaks in the vicinity of adult pines" (p. 643).

The possibility of recognizing an unambiguous "fire adaptation" in a fossil plant is hindered by the fact that many structural features constituting strategies of plant survival and propagation are not uniquely adaptations to fire. For example, the siting of perennating organs and growing points at or below ground level (as for example in the grasses) may be regarded as adaptations to survive seasonal aridity and extremes of cold as well as the effects of wildfire. Nonetheless, there are a few instances in the fossil record where the facts are at least consistent with the existence of fire adaptation. For example, the repeated occurrences of certain Lower Cretaceous ferns in a charred state has led Harris (1981, 56) to suggest their association with a "fire ecology." He is guarded in his conclusions but, referring to a specific Wealden habitat in southern England, favors the occurrence of "fire often enough to keep woody gymnosperms from growing up and suppressing the herbs." In review, we have little reason to doubt that in the past, as now, in any environments in which wildfire occurred it was a powerful selective force influencing the evolution of plant structure and behavior.

Interaction of Wildfire with Other Phenomena

In this brief review we have set out to show that wildfire is a physical phenomenon occupying a central position linking a range of physical and biological processes (fig. 9.3). For reasons considered above we regard climate as the single most important controlling factor governing the potential occurrence of wildfire (fig. 9.3, nos. 1 and 5). Climatic control of wildfire is not only direct, in producing potential fuel and an ignition source (nos. 2 and 6), but indirect, in determining the status of plant communities (nos. 3 and 4) and hence the likelihood of wildfire. Climate has evidently been influenced in the course of earth history by global features, such as latitude (and hence by change of latitude brought about by the movement of continental plates), as well as by altitude, aspect, and other features of the topography (no. 1). Equally, climate may be influenced in a less immediately evident way by volcanic activity through the discharge of ash into the atmosphere (Lamb, 1972). Extraterrestrial phenomena (solar radiation, the changing inclination of the earth's axis of rotation to the ecliptic) are further important factors in paleoclimatic change (Harland, 1981). But the influence of climate on vegetation and hence on the incidence of fire is not a simple direct process. As Kemp (1981) points out, there is an interaction of climate,

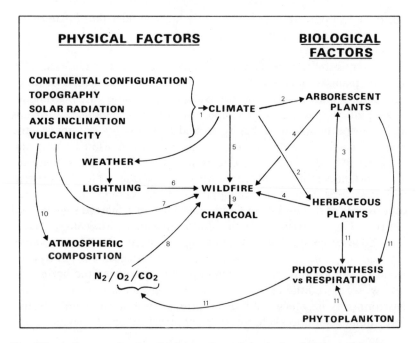

Fig. 9.3. A diagram showing the interaction of physical and biological factors associated with the formation of charcoal by wildfire. For detailed comments see text.

vegetation, and fire that would tend to accelerate any externally induced climate change from a moister to a more arid climatic regime. Considering the response of vegetation to climatic change in Australia during the Tertiary period, Kemp writes:

The general trend through time from rainforest to open sclerophyll vegetation in inland areas has usually been interpreted as a direct result of climatic change in the direction of increasing aridity. It is possible however that natural fires might have accelerated the change towards dry open woodland. Under conditions where arid phases became more common a higher frequency of fires would have favoured the spread of sclerophyllous woodland at the expense of more fire-sensitive rainforest trees.... More frequent fires encourage vegetation that is yet more fire prone, and the trend continues in the direction of further forest degradation. [p. 14]

The role of the composition of the atmosphere, considered above, is an independent but related link between physical and biological processes (fig. 9.3, nos. 8 and 10). The balance of atmospheric oxygen is controlled by photosynthesis and respiration, not only by land plants and animals but by phytoplankton (no. 11). This balance, apart from its central role in sustaining terrestrial life, also influences the potential occurrence of combustion and plant material and hence of charcoal formation by wildfire (nos. 8 and 9). However, as acknowledged above, vulcanicity represents a further probably rare source of ignition (fig. 9.3, no. 7) of wildfire. The occurrence of charcoal through the geological record constitutes the single most important tangible source of evidence for this interaction of physical and biological processes in time past.

References

Alvin, K. L. 1974. Leaf anatomy of *Weichselia* based on fusainised material. *Palaeontology* 17:587–98.

Alvin, K. L., C. J. Fraser, and R. A. Spicer. 1981. Anatomy and palaeoecology of *Pseudofrenelopsis* and associated conifers in the English Wealden. *Palaeontology* 24:759–78.

Ashton, D. H. 1981. Fire in tall open-forests (wet sclerophyll forests). In *Fire and the Australian biota*, ed. A. M. Gill, R. H. Groves, and I. R. Noble, 339–66. Canberra: Australian Academy of Science.

Beall, F. C., and J. W. Eickner. 1970. *Thermal degradation of wood components*. United States Department of Agriculture Forest Research Paper FPL 130.

Beck, C. B., K. Coy, and R. Schmid. 1982. Observations on the fine structure of *Callixylon* wood. *Am. J. Bot.* 69:54–76.

Bradstock, R. 1981. Our phoenix flora. *Aust. Nat. Hist.* 20:223–26.

Brown, A. A., and K. P. Davis. 1973. *Forest fire: Control and use*. New York: McGraw-Hill.

Clark, F. R. S., and D. A. Russell. 1981. Fossil charcoal and the palaeoatmosphere. *Nature* 290:428.

Cohen, A. D., and W. Spackman. 1977. Phytogenic organic sediments and sedimentary environments in the Everglades-mangrove complex. Part II. The origin, description and classification of the peats of southern Florida. *Palaeontographica Abt. B* 162:71–114.

Cope, M. J. 1980. Physical and chemical properties of coalified and charcoalified phytoclasts from some British Mesozoic sediments: An organic geochemical approach to palaeobotany. In *Advances in organic geochemistry, 1979*, ed. A. G. Douglas and J. R. Maxwell, 663–77. Oxford: Pergamon Press.

———. 1981. Products of natural burning as a component of the dispersed organic matter of sedimentary rocks. In *Organic maturation studies and fossil fuel exploration*, ed. J. Brooks, 89–109. London: Academic Press.

Cope, M. J., and W. G. Chaloner. 1980. Fossil charcoal as evidence of past atmospheric composition. *Nature* 283:647–49.

———. 1981. Reply to: "Fossil charcoal and the palaeoatmosphere," [by] F. R. S. Clark and D. A. Russell. *Nature* 290:428.

Coward, D. H. F., and G. W. Jones. 1952. *Limits of flammability of gases and vapours.* U.S. Bureau of Mines Bulletin 503.

Edwards, D. 1982. Fragmentary non-vascular plant microfossils from the late Silurian of Wales. *Bot. J. Linn. Soc.* 84:223–56.

Falcon, R. M. S. 1978. Coal in South Africa. Part III. Summary and Proposals — The fundamental approach to the characterization and nationalization of South Africa's coal. *Miner. Sci. Eng.* 10:130–53.

Friis, E. M., and A. Skarby. 1981. Structurally preserved angiosperm flowers from the Upper Cretaceous of southern Sweden. *Nature* 291:485–86.

———. 1982. *Scandianthus* gen. nov., angiosperm flowers of Saxifragalean affinity from the Upper Cretaceous of southern Sweden. *Ann. Bot.* 50:569–83.

Gill, A. M., R. H. Groves, and I. R. Noble, eds. 1981. *Fire and the Australian biota.* Canberra: Australian Academy of Science.

Grebe, H. 1953. Beziehungen zwischen fusitlagen und pollenfuerung in der rheinischen braunkohle. *Paläont. z.* 27:12–15.

Hare, R. C. 1961. Heat effects on living plants. *U.S. Dep. Agric., Southeast. For. Exp. Stn., Occ. Pap.* 183:1–22.

Harland, W. B. 1981. Chronology of earth's glacial and tectonic record. *J. Geol. Soc., London* 138:197–203.

Harris, T. M. 1957. A Liasso-Rhaetic flora in South Wales. *Proc. R. Soc. London, Ser. B* 147:289–308.

———. 1958. Forest fire in the Mesozoic. *J. Ecol.* 46:447–53.

———. 1981. Burnt ferns from the English Wealden. *Proc. Geol. Assoc.* 92:47–58.

Kemp, E. M. 1981. Pre-Quaternary fire in Australia. In *Fire and the Australian biota,* ed. A. M. Gill, R. H. Groves, and I. R. Noble, 3–21. Canberra: Australian Academy of Science.

Komarek, E. V. 1972. Ancient fires. *Proc. Annu. Tall Timbers Fire Ecol. Conf.* 12:219–40.

Lamb, H. H. 1972. *Climate; Present, past and future.* Vol. 1. London: Methuen.

Lockwood, J. P., and I. S. Williams. 1978. Lava trees and tree moulds as indicators of lava flow direction. *Geol. Mag.* 115:69–74.

McGinnes, A. E., S. A. Kandeel, and P. S. Szopa. 1971. Some structural changes observed in the transformation of wood into charcoal. *Wood and Fibre* 3:77–83.

McGinnes, A. E., P. S. Szopa, and J. E. Phelps. 1974. Use of scanning electron microscopy in studies of wood charcoal formation. 476. *Scanning Electron Microscopy/1974,* pt. 2, pp. 469–76. IIT Research Institute. Chicago.

Moore, P. D. 1978. Forest fires. *Nature* 272:754.

———. 1982. Fire: Catastrophic or creative? *Impact Sci. Soc.* 32:5–14.

Muir, M. 1970. A new approach to the study of fossil wood. *Scanning Electron Microscopy/1970,* pp. 131–36. IIT Research Institute, Chicago.

Raup, H. M. 1981. Physical disturbance in the life of plants. In *Biotic crises in ecological and evolutionary time,* ed. M. H. Nitecki, 39–52. New York: Academic Press.

Raven, P. H., R. F. Evert, and H. Curtis. 1981. *Biology of plants*. 3d ed. New York: Worth.

Rundle, P. W. 1981. Fire as an ecological factor. In *Physiological plant ecology 1. Responses to the physical environment*, ed. O. L. Lange et al., 501–38. Berlin: Springer-Verlag.

Schopf, J. M. 1948. Variable coalification: The processes involved in coal formation. *Econ. Geol.* 43:207–25.

———. 1952. Was decay important in the origin of coal? *J. Sediment. Petrol.* 22:61–69.

———. 1975. Modes of fossil preservation. *Rev. Palaeobot. Palynol.* 20:27–53.

Scott, A. C. 1974. The earliest conifer. *Nature* 251:707–08.

———. 1980. The ecology of some Upper Palaeozoic floras. In *The terrestrial environment and the origin of land vertebrates*, ed. A. L. Panchin, 87–115. Systematics Association Special Volume 15. London: Academic Press.

Scott, A. C., and W. G. Chaloner. 1983. The earliest conifer from the Westphalian B of Yorkshire. *Proc. R. Soc. London, Ser. B* 220:163–82.

Scott, A. C., and M. E. Collinson. 1978. Organic sedimentary particles: Results from scanning electron microscope studies of fragmentary plant material. In *Scanning electron microscopy in the study of sediments*, ed. W. B. Whalley, 137–67. Norwich, CT: Geo Abstracts.

Singh, G., A. P. Kershaw, and Robin Clark. 1981. Quaternary vegetation and fire history in Australia. In *Fire and the Australian biota*, ed. A. M. Gill, R. H. Groves, and I. R. Noble, 23–54. Canberra: Australian Academy of Science.

Summerhayes, C. P. 1981. Organic facies of Middle Cretaceous black shales in deep North Atlantic. *Am. Assoc. Pet. Geol. Bull.* 65:2364–80.

Swain, T., and G. Cooper-Driver. 1981. Biochemical evolution in early land plants. In *Palaeobotany, palaeoecology and evolution*, ed. K. J. Niklas, 103–34. New York: Praeger.

Swanson, F. J. 1981. Fire and geomorphological processes. In *Fire regimes and ecosystem properties*, ed. H. A. Mooney et al., 401–20. U.S. Forest Service General Technical Report WO-26.

Tappan, H. 1968. Primary production, isotopes, extinctions and the atmosphere. *Palaeogeogr., Palaeoclimatol., Palaeoecol.* 4:187–210.

Teichmuller, M. 1975. Origin of the petrographic constituents of coal. In *Stach's textbook of coal petrology*, ed. E. Stach et al., 176–238. Berlin: Gerbrüder Borntraeger.

Terasmae, J., and N. C. Weeks. 1979. Natural fires as an index of paleoclimate. *Can. Field Naturalist* 93:116–25.

White, D., and R. Thiessen. 1913. *The origin of coal*. U.S. Bureau of Mines Bulletin 38.

Williamson, G. B., and E. M. Black. 1981. High temperature of forest fires under pines as a selective advantage over oaks. *Nature* 293:643–44.

Index

abiotic environments: of Pennsylvanian, 250

Acacus Formation (Libya): age of, 58

Acadia: Devonian position, 103; Early Carboniferous paleoflora, 180, 187–88, 194, 195, 196, 197, 208–09; Early Carboniferous character, 194–95, 208–09; Early Carboniferous paleoclimate, 195

Acadian mountains: Devonian effect, of, 113

acanthodians: appearance of, 63–64, 65; Gedinnian-Emsian, 66; Eifelian-Famennian, 71

acarids: Early Devonian, 67

Acercostraca: in Early Devonian, 67

Achlamydocarpon varius, 234

acritarchs: possible eukaryotes, 14

actinopterygians: appearance in Devonian, 71; replace agnathans in Devonian, 72

adaptations: permanent, 24; in early macrophytes, 78–79, 81, 84

adaptive zone: of Pennsylvanian species, 237, 250

Adiantites, 188, 193

Aegir Horizon, Netherlands, 238

aerobic metabolism, 31

Africa: Devonian paleofloras, 151; Early Carboniferous paleoflora of, 171, 187, 188, 207, 209; Ordovician trilobites of, 175; Early Carboniferous paleoclimate, 201

aglaspid, 49

agnathan fish: Ordovician, 49; Silurian, 60, 61, 63; Devonian, 71, 72

albedo: pre-Silurian, 53; considered in modeling, 114

Alethopteris, 240

algae: multicellular, 17; terrestrial adaptations of, 24; distinctions from higher plants, 27; ancestral to land plants, 28; physiology, 31, 36; UV screen in, 33; pre-Silurian, 49; Late Silurian, 63; in Devonian terrestrial food chain, 67

algal and lichen crusts: effect on weathering, 50, 51

algal debris: as a wildfire fuel source, 267

allochthonous terranes (microplates): in Devonian reconstructions, 103

Amazonas (basin): effect on Gondwanan climate, 109

amphiaerobic organisms, 31

amphibians: Devonian terrestrial, 71–72

anaerobic metabolism, 31

anaspids: Late Silurian, 60, 63; feeding habits, 65; Gedinnian-Emsian, 66

Aneimites, 180, 188, 190

Angara: Siegenian-Emsian paleoflora, 153; Early Carboniferous paleoflora, 170, 171, 180, 181, 192; Early Carboniferous paleoclimate, 206

Angaropteridium, 171

angiosperms: environmental influence of, 7–8; influence on marine sphere, 8; chemistry of, 8

annelids: in Devonian food chain, 71